The Craft of Research

Chicago Guides
to *Writing*, Editing,
and Publishing

On Writing, Editing, and Publishing
JACQUES BARZUN

Tricks of the Trade
HOWARD S. BECKER

Writing for Social Scientists
HOWARD S. BECKER

The Craft of Translation
JOHN BIGUENET AND RAINER SCHULTE, EDITORS

The Craft of Research
WAYNE C. BOOTH, GREGORY G. COLOMB, AND JOSEPH M.
WILLIAMS

Glossary of Typesetting Terms
RICHARD ECKERSLEY, RICHARD ANGSTADT, CHARLES M.
ELLERSTON, RICHARD HENDEL, NAOMI B. PASCAL, AND ANITA
WALKER SCOTT

Writing Ethnographic Fieldnotes
ROBERT M. EMERSON, RACHEL I. FRETZ, AND LINDA L. SHAW

Legal Writing in Plain English
BRYAN A. GARNER

Getting It Published
WILLIAM GERMANO

A Poet's Guide to Poetry
MARY KINZIE

Mapping It Out
MARK MONMONIER

The Chicago Guide to Communicating Science
SCOTT L. MONTGOMERY

Indexing Books
NANCY C. MULVANY

Getting into Print
WALTER W. POWELL

A Manual for Writers of Term Papers, Theses, and Dissertations
KATE L. TURABIAN

Tales of the Field
JOHN VAN MAANEN

Style
JOSEPH M. WILLIAMS

A Handbook of Biological Illustration
FRANCES W. ZWEIFEL

Chicago Guide for Preparing Electronic Manuscripts
PREPARED BY THE STAFF OF THE UNIVERSITY OF CHICAGO PRESS

The Craft of Research

SECOND EDITION

WAYNE C. BOOTH

GREGORY G. COLOMB

JOSEPH M. WILLIAMS

THE UNIVERSITY OF CHICAGO PRESS

Chicago & London

WAYNE C. BOOTH is the George Pullman Distinguished Service Professor Emeritus at the University of Chicago. His many books include *The Rhetoric of Fiction* and *For the Love of It: Amateuring and Its Rivals,* both published by the University of Chicago Press.

GREGORY G. COLOMB is professor of English language and literature at the University of Virginia. He is the author of *Designs on Truth: The Poetics of the Augustan Mock-Epic.*

JOSEPH M. WILLIAMS is professor emeritus in the Department of English Language and Literature at the University of Chicago. He is the author of *Style: Toward Clarity and Grace.* Together Colomb and Williams have written *The Craft of Argument,* published by the University of Chicago Press.

The University of Chicago Press, Chicago 60637
The University of Chicago Press, Ltd., London

© 1995, 2003 by The University of Chicago
All rights reserved. Published 2003
Printed in the United States of America

12 11 10 09 08 07 06 05 04 03   1 2 3 4 5
ISBN: 0-226-06567-7 (cloth)
ISBN: 0-226-06568-5 (paper)

Library of Congress Cataloging-in-Publication Data
Booth, Wayne C.
   The craft of research / Wayne C. Booth, Gregory G. Colomb, Joseph M. Williams.—2nd ed.
        p. cm. — (Chicago guides to writing, editing, and publishing)
Includes bibliographical references and index.
     ISBN 0-226-06567-7 (cloth : alk. paper) — ISBN 0-226-06568-5 (paper : alk. paper)
     1. Research—Methodology. 2. Technical writing. I. Colomb, Gregory G. II. Williams, Joseph M. III. Title. IV. Series.
    Q180.55.M4 B66 2003
    001.4'2—dc21

2002015184

Contents

Preface

We intend that, like the first edition of *The Craft of Research*, this second edition meet the needs of all researchers, not just beginners, or advanced graduate students, but even those in business and government who are assigned research on any topic, technological, political, or commercial. Our aim is to

- guide you through the complexities of organizing and drafting a report that poses a significant problem and offers a convincing solution;

- show you how to read your drafts as your readers might so that you can recognize passages they are likely to find unnecessarily difficult and then revise them effectively.

Other handbooks touch on these matters, but this one differs in many ways. Most current guides agree that researchers never move in a straight line from finding a topic to stating a thesis to filling in note cards to drafting and revision. Real research loops back and forth, moving forward a step or two, going back and moving ahead again, anticipating stages not yet begun. But so far as we know, no previous guide has tried to explain how each part of the process influences all the others—how asking questions about a topic prepares the researcher for drafting, how drafting reveals problems in an argument, how writing an introduction can send you back to the library.

THE COMPLEXITIES OF THE TASK

Because research is so complex, we have tried to be explicit about it, including matters that are usually left implicit as part of a mysterious creative process, including these:

- how to turn a vague interest into a problem worth posing and solving;

- how to build an argument that motivates readers to accept your claim;

- how to anticipate the reservations of thoughtful but critical readers and then respond appropriately;

- how to create an introduction and conclusion that answer that toughest of questions, *So what?*;

- how to read your own writing as others may, and thereby learn when and how to revise it.

Central in every chapter is our advice to side with your readers, to imagine how they judge what you have written. Meeting their expectations is not, however, the only reward for mastering the formal elements of a research report. When you learn those formal matters, you are better able to plan, conduct, and evaluate the process that creates one. The elements of a report—its structure, style, and methods of proof—are not empty formulas for convincing readers to accept your claims. They help you test your work and discover new directions in it.

As you can guess, we believe that the skills of doing and reporting research are not just for the elite; they can be learned by all students. Though some aspects of advanced research can be learned only in the context of a specific community of researchers, the good news is that even if you don't yet belong to such a community, you can create something like it on your own. To that end, in our "Postscript for Teachers," we show you (and your teachers) ways that a class can create such a community.

We should note what we do not address. We do not discuss how to incorporate narratives and "thick descriptions" into an argument. Nor have we examined how arguments incorporate recordings and other audio forms of evidence. Both are important issues, but too large for us to do justice to them here. There are also advanced techniques for Internet searches and other ways of gathering data that we do not have space to cover. Our bibliography suggests a number of sources for guidance in those areas.

ON THE SECOND EDITION

In revising the first edition, we have naturally been grateful to all those who praised it, but especially to those who used it. We hoped for a wide audience, but didn't expect it to be as wide as it turned out to be, ranging from first-year students in composition classes to advanced graduate students to advanced researchers (including more than a few tenured professors, if we can believe our e-mail). We are particularly thankful to all those users who shared their suggestions for improvement.

Because the reception of the first edition was so positive, we were at first uneasy about doing a second. We didn't want to lose whatever it was that readers of the first found useful. Yet we had learned some things in the last ten years, and we knew the book had places that could be improved. (Besides, the three of us always hope for the chance to do one more draft of everything we write.)

We have cleaned things up in every chapter, cut repetitions, and fixed sentences that were less than felicitous. We have expanded our comments on how computers have changed research. We have extensively revised the chapters on argument to explain a number of issues more clearly. We have also made a crucial distinction that we missed in the first edition—the difference between reasons and evidence. (How we let that one get by, we'll never know; it is small comfort that few if any other books on research arguments make that distinction either.) We have modified what we said about *qualifi-cations* and *rebuttals,* which we now call *acknowledgment* and *response.* We have also redone the chapter on the visual representation of data. Finally, we have rearranged the order of chapters a bit. Throughout, we have tried to preserve the tone, the voice, the sense of directness that so many of you thought was important in the first. We have revised to make things better, but sometimes revisions make them worse. We hope we have made them better.

OUR DEBTS

We want again to thank the many without whose help the first edition could never have been realized, especially Steve Biegel, Jane Andrew, and Donald Freeman. The chapter on the visual presentation of data was improved significantly by the comments of Joe Harmon and Mark Monmonier. We would also like to thank those who helped us select and edit the "Appendix on Finding Sources": Jane Block, Diane Carothers, Tina Chrzastowski, James Donato, Kristine Fowler, Clara Lopez, Bill McClellan, Nancy O'Brien, Kim Steele, David Stern, Ellen Sutton, and Leslie Troutman. We are also indebted to those at the University of Chicago Press who, when we agreed to undertake this project almost a decade ago, kept after us until we finally delivered.

For this second edition, we'd like to thank those whose thoughtful reviews of the first edition and our early revisions of it helped us see opportunities we would otherwise have missed: Don Brenneis, University of California, Santa Cruz; John Cox, Hope College; John Mark Hansen, University of Chicago; Richard Hellie, University of Chicago; Susannah Heschel, Dartmouth

> **A TRUE STORY**
> As we were preparing this second edition, Booth got a call from a former student who, as had all of his students, been directed again and again by Booth to revise his work. Now a professional in his mid-forties, he called to tell Booth about a dream he had had the night before: "You were standing before Saint Peter at the Pearly Gate, hoping for admission. He looked at you, hesitant and dubious, then finally said, 'Sorry, Booth, we need another draft.'"

College; Myron Marty, Drake University; Robert Sampson, University of Chicago; Joshua Scodel, University of Chicago; W. Phillips Shively, University of Minnesota; and Tim Spears, Middlebury College.

We are also grateful to Alec MacDonald and Sam Cha for their invaluable help tracking down details of all sorts, and to Adam Jernigan for his careful reading of the manuscript. All three were quick and reliable.

We are again indebted to those at the University of Chicago Press who supported the writing of this revision.

From WCB: I am amazed as I think back on my more than fifty years of teaching and research by how many students and colleagues could be cited here as having diminished my ignorance. Since that list would be too long, I'll thank mainly my chief critic, my wife, Phyllis, for her many useful suggestions and careful editing. She and my daughters, Katherine Stevens and Alison Booth, and their children, Robin, Emily, and Aaron, along with all those colleagues, have helped me combat my occasional despair about the future of responsible inquiry.

From GGC: I, too, have been blessed with students and colleagues who have taught me much—first among them the hundreds of grad students who shared with me their learning to be teachers. They, above all, have shown me the possibilities in collaborative inquiry. What I lean on most, though, are home and family: Sandra, Robin, Karen, and Lauren. Through turbulent times and calm, they gave point and purpose to it all. Before them was another loving family, whose center, Mary, still sets an example to which I can only aspire.

From JMW: The family has grown since the first edition, and I am ever more grateful for their love and support: Ol, Chris, Dave and Patty, Megan and Phil, Joe and Christine, and now Lily and the twins, Nicholas and Katherine. And at beginning and end, Joan, whose patience, love, and good sense flow still more bountifully than I deserve.

PART I

Research, Researchers, and Readers

Prologue

STARTING A RESEARCH PROJECT

If you are beginning your first research project, the task may seem overwhelming. *How do I find a topic? Where do I find information on it? What do I do with it when I find it?* Even if you have written a research paper in a writing class, the idea of another may be even more intimidating if this time it's supposed to be *the real thing.* Even experienced researchers feel anxious when they tackle a new project, especially when it's of a new kind. So whatever anxiety you may feel, most researchers have felt the same. (It's a feeling that we three know well.) The difference is that experienced researchers know what lies ahead—hard work, but also the pleasure of the hunt; some frustration, but more satisfaction; periods of confusion, but confidence that, in the end, it will all come together.

MAKING PLANS
Experienced researchers also know that research most often comes together when they have a plan, no matter how rough. Before they start, they may not know exactly what they are looking for, but they know in general what they will need, how to find it, and what it should look like when they do. And once they assemble their materials, they don't just start writing, any more than competent builders just start sawing: they make a plan— maybe no more than a sketch of an outline, not even on paper.

3

But shrewd researchers don't let their plan box them in. They change it if they run into a problem or discover in some byway something more interesting that leads in a new direction. But they do start with a plan.

A newspaper reporter, for example, follows a plan when she writes her story as an inverted pyramid, putting the salient information first. But she doesn't do that just for her own benefit, to make her job of drafting easier, but so that *readers* can find the gist of the news quickly, then decide whether to read on. An accountant knows how to plan an audit report, but that plan also lets *investors* quickly find the information they need to decide whether the company is an Intel or another Enron. Within these forms, of course, writers are free to take different points of view, emphasize different ideas, and put a personal stamp on their work. But they also know that when they follow a standard plan, they make it easier for themselves to write and their readers to read efficiently and productively.

The aim of this book is to help you create, execute, and if necessary revise a plan that lets you not only do your own best, most original thinking, but draft a report that meets your readers' needs and their highest expectations.

THE VALUE OF RESEARCH

But first a candid question: Why do research at all? Aside from a grade, what's in it for you?

For those new to research, there are immediate and practical benefits. Learning to do research will help you understand the material you cover as no other kind of work can match. You can evaluate what you read most thoughtfully only when you have experienced the uncertain and often messy process of doing your own research. Writing a report of your own will help you understand the kind of work that lies behind what you find in your textbooks and what experts tell the public. It lets you experience firsthand how new knowledge depends on which questions are asked and which aren't; how the standard forms for presenting research shape the kinds of questions you ask and answers you offer.

More distantly, the skills you learn now will be crucial when you do advanced work in whatever field you choose to study. Even more distantly, the skills of research will pay off long after you leave school, especially in a time aptly named the "Age of Information" (or, too often, of *Mis*information). Sound research reported clearly has immense value now that the Internet and cable flood us with more information than we can absorb, much less evaluate, especially when so much of it is based on research that we rely on at our peril. And though some might think it idealistic, a final reason for doing research is the pleasure it offers in solving a puzzle, the satisfaction of discovering something that no one else knows and that contributes to the wealth of human knowledge and understanding.

Research, though, is not the sort of thing you learn once and for all. Each of the three of us has faced research projects that forced us to take a fresh look at how we do our work. Whenever we've addressed our research to a new research community, we've had to learn its principles to help us focus on what is important to its members. But even then, we could still rely on some common principles that all research communities follow, principles that we describe in this book. We think these principles will be useful not only now but through the years as your circumstances change and your research assignments (and your readers) become increasingly demanding.

But we must be candid: Doing research carefully and reporting it clearly are hard work. They consist of many tasks, often competing for your attention at the same time. However carefully you plan, research follows a crooked path, taking unexpected turns, even looping back on itself. As complex as that process is, though, we will work through it step-by-step. When you can manage the parts, you can manage the whole and then look forward to more research with greater confidence.

FLOODS OF MISINFORMATION

Since the 9/11 attack on the World Trade Center, the U.S. government has been challenged not only to root out terrorism but to counter bizarre claims that have circulated in the Middle East, especially on the Internet: no Muslims were among the hijackers; Jews had advance notice and stayed home; the attacks were the work of the CIA. These claims have been widely believed to be true, even though no evidence backs them up.

Before we feel superior, however, we should recall some bizarre stories believed by many Americans: the CIA started the AIDS epidemic to kill homosexuals and African Americans; the government still hides the bodies of aliens in Area 51; bar codes are a UN conspiracy to take over the world. Every society succumbs to outlandish beliefs, but we all can learn to see through them and to make a case for what we believe is true. It won't convince everyone, but it might convince some, including ourselves.

HOW TO USE THIS BOOK

The best way to deal with the complexity of research (and its anxiety) is to read this book twice. First skim it to see what lies before you (skip ahead when you feel confused or bored). Then as you begin your work, read carefully the chapters relevant to the task at hand. If you are wholly new to research, start rereading from the beginning. If you are in an intermediate course but not yet at home in your field, skim part I, then concentrate on the rest. If you consider yourself an experienced researcher, you will probably find chapter 4 and parts III and IV most useful.

In part I, we address what those of you undertaking your first project have to think about consciously: why readers expect you to write up your research in particular ways (chapter 1), and why you should think of your project not as solitary work but as a conversation with those whose work you read and then with those who will read yours (chapter 2).

In part II, we discuss how to frame and develop your project. We explain

- how to carve out a topic from an interest, then how to focus and question it (chapter 3);

- how to transform those questions into a research problem (chapter 4);

- how to find sources to guide the search for answers (chapter 5);

- how to use those sources and think through what you find (chapter 6).

In part III, we discuss how to assemble a sound case in support of your claim. That includes

- an overview of the elements of a research argument (chapter 7);

- what counts as a significant claim (chapter 8);

- what count as good reasons and sound evidence (chapter 9);

- why and how you must acknowledge questions, objections, and alternatives and respond to them (chapter 10);

- how you justify the logic of your argument (chapter 11).

In part IV, we lay out the steps in producing your report:

- how to plan and write a first draft (chapter 12);

- how to test and revise it (chapter 13);

- how to write an introduction and conclusion that convince readers that your report is worth their time (chapter 14);

- how to present complex quantitative evidence clearly and pointedly (chapter 15);

- how to edit your style to make it clear, direct, and readable (chapter 16).

In an afterword, "The Ethics of Research," we reflect on a matter that goes beyond professional competence. Doing and reporting research is a social activity with an ethical dimension. We all know of recent scandals about the dishonest research of historians, scientists, stock analysts, and others, and we see plagiarism spreading among writers at all levels of achievement, from secondary school students to those at the top of their professions. Such events emphasize the importance of hard thinking about what constitutes *ethical* research and its reporting.

Between some of the chapters you will find "Quick Tips," brief sections that complement the chapters. Some Quick Tips are checklists; some discuss additional considerations for advanced students; several address matters not raised in the chapters. But all add something new.

At the end of this book, there is a brief survey of recent work in the issues we address in this book, an essay aimed at those who teach research, and a bibliography of sources for beginning researchers and for those getting into particular fields.

Research is hard work, but like any challenging job done well, both the process and the results can bring real personal satisfaction. No small part of that satisfaction comes from knowing that your work supports and sustains the fabric of your community. That sense of contributing to a community is never more rewarding than when you discover something that you believe can improve your readers' lives by changing what and how they think.

CHAPTER ONE

Thinking in Print

THE USES OF RESEARCH, PUBLIC AND PRIVATE

In this chapter, we define research, then discuss how you will benefit from learning to do it well, why we value it, and why we hope you will learn to value it too.

Whenever you read about a scientific breakthrough or a crisis in world affairs, you benefit from the research of those who reported it, who themselves benefited from the research of countless others. When you stand in the reading room of a library to pursue your own work, you are surrounded by centuries of research. When you log on to the Internet, you have access to millions of research reports. All those reports are the product of researchers who have posed endless questions and problems, gathered untold amounts of information, worked out answers and solutions, and then shared them with the rest of us.

Teachers at all levels devote their lives to research. Governments spend billions on it, and businesses even more. Research goes on in laboratories and libraries, in jungles and ocean depths, in caves and in outer space. It stands behind every new technology, product, or scientific discovery—and most of the old ones. Research is in fact the world's biggest industry. Those who cannot reliably do research or evaluate the research of others will find themselves on the sidelines in a world that increasingly depends on sound ideas based on good information produced by trustworthy inquiry.

In fact, research reported by others, in writing, is the source of most of what we all believe. Of your three authors, only Williams has ever set foot in Australia, but Booth and Colomb are certain that it exists, because for a lifetime they have read about it in reports they trust and seen it on reliable maps (and heard about it from Williams). None of us has been to Venus, but we believe that it is hot, dry, and mountainous. Why? Because that's what we've read in reports we trust. Whenever we "look something up," our research depends on the research of others. But we can trust their research only if we can trust that they did it carefully and reported it accurately.

1.1 WHAT IS RESEARCH?

In the broadest terms, everyone does research: we all gather information to answer a question that solves a problem. You do it every day.

> PROBLEM: You need a new head gasket for a '65 Mustang.
> RESEARCH: You call auto parts stores or get on the Internet to see who has one in stock.

> PROBLEM: You want to know where Michael Jordan was born.
> RESEARCH: You go to the library and look in a biographical dictionary. Or you call up Google.com and then sort through the 410,000+ references to him.

> PROBLEM: You want to learn more about a discovery of a new species of tropical fish.
> RESEARCH: You search the Internet for articles in newspapers or magazines.

Though we all do that kind of research, we don't all write it up. But we do rely on those who did: the auto parts suppliers, Jordan's biographers, and the fish discoverers—all wrote up the results of their research because they anticipated that one day someone would have a question that their data would answer.

In fact, without trustworthy and tested *published* research available to all of us, we would be locked in the opinions of the moment, either prisoners of what we alone experience or dupes to everything we hear. Of course, we all want to believe that our opinions are sound; yet mistaken ideas, even dangerous ones, flourish because too many people accept too many opinions on not very good evidence. And those who act on unsound opinions can lead themselves, and others, to disaster. Just ask the thousands who invested in the failed energy giant Enron because they heard so many good opinions of it from analysts and the media. Only after Enron's deceptive bookkeeping was exposed and analyzed in writing did they see how those high opinions were based on bad, sometimes even faked research.

That's why in this book we will urge you to be amiably skeptical of most of the research you read, to question it, even as you realize how thoroughly you depend on it. Are we three authors 100 percent drop-dead certain that reports of Venus being hot, dry, and mountainous are true? No, but we trust the researchers who have published reports about it, as well as the editors, reviewers, and skeptical readers who have tested those reports and published their own results. So we'll go on thinking that Venus is hot and dry until other researchers report better evidence, tested by other researchers, that shows us otherwise.

If you are reading this book because a teacher has assigned you a research project, you might be tempted to treat it as just a chore or an empty exercise. We hope you won't. You have practical reasons to take the work seriously: you will learn skills that pay off in almost any career you choose. Beyond that, your project invites you to join the oldest and most esteemed of human conversations, one that has been conducted for millennia among philosophers, engineers, biologists, social scientists, historians, literary critics, linguists, theologians—the list of researchers is endless.

Right now, you may feel that the conversation seems one-sided, that you have to listen more than you can speak, and that in any event you have little to contribute. That may be true for the moment. But at some point you will be asked to join a conversation that, at its best, can help you and your community free yourselves from ignorance, prejudice, misunderstanding, and the half-baked ideas that so many charlatans try to impose on us. The world changes every day because of research, not always for the better. But done well, research is crucial to improving every facet of our lives. It is no exaggeration to say that your research and your reports of it can improve perhaps not the whole world, but at least your corner of it.

1.2 WHY WRITE IT UP?

For some of you, though, the invitation to join the conversation of research may still seem easy to decline. If you undertake it, you will face demanding tasks in finding a good question, searching for sound data, finding and supporting a good answer, and then writing it all up. Even if you turn out a first-rate report, it will likely be read not by an eager world, but only by your teacher. *And, besides,* you may think, *my teacher knows all about my topic. If she just told me the answers or pointed me to the right books, I could concentrate on learning what's in them. What do I gain from writing up my research, other than proving I can do it?*

Here are some answers.

1.2.1 Write to Remember

Researchers write up what they find just to remember it. A few lucky people can retain information without recording it, but most of us get lost when we think about what Smith found in light of Wong's position, and compare both to the odd data in Brunelli, especially as they are supported by Boskowitz—*But wait a minute. I've forgotten what Smith said!* Most researchers can plan and conduct their project only with the help of writing—by listing sources, assembling research summaries, keeping lab notes, making outlines, and so on. What you don't write down you are likely to forget or, worse, to misremember. That's why careful researchers don't wait until they've gathered all their data to start writing: they write from the beginning of their project so that they can hold as much of it in their minds as clearly as they can.

1.2.2 Write to Understand

A second reason for writing is to understand. When you arrange and rearrange the results of your research in new ways, you discover new connections, contrasts, complications, and implications. Even if you could hold in mind everything you found, you would need help to line up arguments that pull in different directions, plot out complicated relationships, sort out disagreements among experts. *I want to use these claims from Wong, but her argument is undercut by Smith's data. When I compare them, I see that Smith ignores this last part of Wong's argument. Aha! If I introduce it with this part from Brunelli, I can focus on the part of Wong's argument that lets me question Smith.* Writing supports thinking, not just by helping you understand better what you have found, but by helping you find in it larger patterns of meaning.

1.2.3 Write to Gain Perspective

The basic reason for writing, though, is to get your thoughts out of your head and onto paper, where you can see them in the clearer light of print, a light that is always brighter and usually less flattering. Just about all of us, students and professionals alike, think our ideas are more coherent in the dark warmth of our minds than they turn out to be in the cold light of day. You improve your thinking when you encourage it with notes, outlines, summaries, commentary, and other forms of thinking on paper. But you can't know what you really *can* think until you separate specific ideas from the swift and muddy flow of thought and fix them in an organized, coherent form.

In short, you should write so that you can remember more accurately, understand better, and see what you think more clearly. (And as you will discover, the better you write, the more critically you will read.)

1.3 WHY A FORMAL REPORT?

Even if you agree that writing is an important part of learning, thinking, and understanding, some of you may still wonder why you can't write it your own way, why you must satisfy the formal constraints imposed by a research community, particularly one that you may not yet belong to (or even want to). The constraints imposed by writing for others often vex students who believe they have no reason to conform to the practices of a conversation they did nothing to create. *I don't see why I should adopt language and forms that are not mine. What's wrong with my own language? Aren't you just trying to turn me into an academic like yourself? If I write as my teachers expect me to, I risk losing my own identity.*

Such concerns are legitimate (students should raise them more often). But it would be a feeble education that did not change you at all, and the deeper your education, the more it will change the "you" that you think you are, or want to be. That's why it is so important to choose carefully what you study and with whom. But it would be a mistake to think that learning to write sound research reports must threaten your true identity. Learning to do research will not turn you into a clone of your teachers. It will change the way you think, but only by giving you more ways of thinking. You may be different, but you will also be freer to choose who you want to be and what you want to do next.

Perhaps the most important reason for learning to report research in ways readers expect is that you learn more about your ideas and about yourself by testing them against the standards and values of others. Writing for others demands more from you than writing for yourself. By the time you fix your ideas in writing, they are so familiar to you that you need help to see them not for what you want them to be but for what they really are. You reach that end only by imagining, and then meeting, the needs and expectations of others: you create a kind of transaction between you and your readers—what we like to call a *rhetorical community*.

That's why traditional forms and plans are more than empty vessels into which you pour your findings. Those forms have evolved to help writers see their ideas in the brighter light of their readers' expectations and understanding. You will understand your own work better when you explicitly try to anticipate your readers' questions: *How have you evaluated your evidence? Why do you think it is relevant? How do your claims add up? What ideas have you considered but rejected? How can you respond to your readers' predictable questions, reservations, and objections?* All researchers can recall a moment when writing to meet their readers' expectations revealed a flaw or a blunder, or even a great opportunity that escaped them in a first draft written for themselves.

Traditional forms embody the shared practices and values of a research community, matters that contribute to the identity not only of that community but of each of its members. Whatever community you join, you'll be expected to show that you understand its practices by reporting your research in ways that have evolved to communicate it. Once you know the standard forms, you'll have a better idea about your particular community's predictable questions and understand better what its members care about, and why. But what counts as good work is the same in all of them, regardless of whether it is in the academic world or the world of government, commerce, or technology. If you learn to do research well now, you gain an immense advantage, regardless of the kind of research you will do later.

1.4 CONCLUSION

Writing a research report is, finally, thinking in print, but thinking from the point of view of your readers. When you write with others in mind, you give your ideas the critical attention they need and deserve. You disentangle them from your memories and wishes, so that you—and others—can explore, expand, combine, and understand them more fully. Thinking in written form for others can be more careful, more sustained, more attuned to those with different views—more thoughtful—than just about any other kind of thinking.

You can, of course, choose the less demanding path: do just enough to satisfy your teacher. This book can help you do that. But you will shortchange yourself if you do. If instead you find a topic that *you* care about, ask a question that *you* want to answer, your project can have the fascination of a mystery whose solution rewards your efforts in finding it. Nothing contributes more to a successful research project than your commitment to it.

We wish we could tell you how to balance your belief in the worth of your project with the need to accommodate the demands of teachers and colleagues, but we cannot. If you believe in what you're doing and cannot find anyone else who shares your belief, all you can do is put your head down and press on. With our admiration.

> Some of the world's most important research has been done by those who persevered in the face of indifference or even hostility, because they never lost faith in their vision. The geneticist Barbara McClintock struggled for years unappreciated because her research community considered her work uninteresting. But she believed in it and pressed on. When her colleagues finally realized that she had already answered questions that they were just starting to ask, she won science's highest honor, the Nobel Prize.

CHAPTER TWO

Connecting with Your Reader

(RE)CREATING YOUR SELF AND YOUR AUDIENCE

Your research counts for little if no one reads it. Yet even experienced researchers sometimes forget to keep their readers in mind as they plan and draft. In this chapter we show you how to think about readers as you begin your research. We also explain one of the best ways to anticipate how readers will respond—working in collaboration with others.

Most of the important things we do, we do with others. Some students think that research is different: they imagine a solitary scholar reading alone in a hushed library or peering into a microscope surrounded only by glassware and computers. But no place is more filled with voices than a library or lab. Even when you work alone, you silently converse with others when you read a book or call up a website. Every time you go to a source for information, you renew a relationship between writers and readers that may be centuries old. And when you report your own research, you can hope that other voices will respond to yours, so that you can in turn respond to them. And so it goes.

But conversation is a social activity. Both sides have to understand what each expects of the other, what "social role" each is expected to play. And that's especially true when the conversation is in writing and among professional colleagues.

2.1 CREATING ROLES FOR WRITERS AND READERS

When we talk with others in person, we judge them by how well they play the roles expected of them: do they listen carefully, make claims thoughtfully, answer questions directly? It's the same when you read: *Hmmm, Abrams is modest but not careful about this evidence. Quincy has good data but overgeneralizes.* (Right now, we three expect that you are judging us.) But just as in conversation, these judgments go both ways: readers judge a writer, but a thoughtful writer has in advance also judged her readers, by imagining who they are, what they are like, what they know, what they need and want. And then she uses that judgment to shape what she writes.

For example, the writer of these next two passages judged that she was addressing readers with different levels of knowledge about the chemistry of heart muscles. So she imagined herself in very different relationships with them:

> 1a. The control of cardiac irregularity by calcium blockers can best be explained through an understanding of the calcium activation of muscle groups. The regulatory proteins actin, myosin, tropomyosin, and troponin make up the sarcomere, the basic unit of muscle contraction.

> 1b. Cardiac irregularity occurs when the heart muscle contracts uncontrollably. When a muscle contracts, it uses calcium, so we can control cardiac irregularity with drugs called calcium blockers. To understand how they work, it is first necessary to understand how calcium influences muscle contraction. The basic unit of muscle contraction is the sarcomere. It consists of four proteins that regulate contraction: they are actin, myosin, tropomyosin, and troponin.

In (1a) the writer seems to cast herself and her readers in the roles of equally knowledgeable expert colleagues; in (1b) she casts her reader as someone who knows nothing about the subject and herself as the patient expert, slowly explaining a complicated issue. If she judged correctly, her readers will judge her favorably. But when a writer miscasts readers, she can lose their trust and often their willingness to read. Had she switched audiences for those passages, the nonexpert would likely think (1a) indifferent to his needs and her expert colleagues would judge (1b) to be condescendingly simpleminded.

In fact, writers cannot avoid creating a role for their readers. That's why, in writing this book, we tried to imagine you—what you're like, what you know about research, whether you even care about it. We cast you in a role, created a *persona* for you that we hoped you would comfortably adopt. Then we imagined ourselves in our own persona, talking to the "you" that we imagined you would be willing to be. That was not easy, because there are so many "you's" out there, all different. We hoped to speak as comfortably to those of you starting your first serious research project as to those well into your careers. Only you can judge how well we've managed to talk to and *with* all of you.

These personas and the relationship you create with your own readers are so important that they are worth thinking about well before you envision a first draft. If you miscast readers, your mistake will leave in your early drafts so many traces that you won't easily fix them in the final one.

2.2 CREATING A RELATIONSHIP WITH YOUR READER: YOUR ROLE

Few people read research reports just for fun. So you have to know what you can offer readers to create a relationship that makes them want to read your report. Beginning researchers too often offer a relationship that caricatures a bad classroom exchange: *Teacher, I know so much less than you, who will give me a grade. So my role is to show you how much information I dug up, and yours is to decide whether I've found enough.* That's a big mistake. Not only does it demean both you and your teacher, but it makes your project just one long, pointless drill. Worst of all, it casts you in a role exactly opposite to that of a true researcher.

In a research report, you have to reverse the roles of teacher and student. As a researcher, you have to adopt the role of someone who knows what others need to know and to cast your reader as someone who doesn't know but needs to. That will be easier if you find a research question that you want to answer and your teacher can't, without your help. (In fact, your teacher is likely to know less than you about your specific question.) But even if not, you have to step into the kind of relationship researchers have with their readers, one that goes beyond *Here are the facts I've dug up about medieval Tibetan weaving. Did I get them right?*

So your first step in establishing a sound *research* relationship with readers is to offer them more than a collection of known facts. There are three such offers that experienced researchers typically make; the third is most common in academic research. As you begin, imagine that you will offer and your readers will accept one of the three following relationships, but most likely the third.

2.2.1 I've Found Something Really Interesting

You take a step beyond mere data-grubbing when you can say to your readers, *Let me share some information about medieval Tibetan weaving that I think is really interesting.* If you have learned something that interests *you* and you can demonstrate that interest in your report, that's the best start you can make in learning to do sound research. In an introductory writing course, the interest you seem to take in your work will roughly predict the interest your teacher will take in it.

Ideally, of course, you want her to be as interested in Tibetan weaving as you are, and if you are in a class in Asian art, she may be. But even if not, you still have to cast yourself in the role of someone who has found something interesting, maybe even new and important, *at least to yourself,* and to cast your reader in the role of someone equally interested. As you become more experienced, you'll also be responsible for actually finding an audience who shares those interests. But at the start, you must at least find a role for yourself that shows your own interest, even enthusiasm for what you've found.

2.2.2 I've Found a Solution to a Practical Problem Important to You

You take a bigger step toward focused research when you can imagine saying to readers not just *I have information that might interest you,* but *My information will help you solve a problem you care about.* That is the kind of research that people in business, commerce, and government do every day. They confront problems whose solutions require research, first just to understand them, and then to figure out how to solve them, problems ranging from homelessness to falling profits to terrorism.

To help you learn that role, teachers sometimes invent "real world" scenarios: an environmental science professor might assign you to write a report for the director of the state Environmental Protection Agency on what to do about cleaning up toxins in a local lake. In this scenario you are not a student dumping data on a teacher, but someone who must play the role of a scientist giving practical, pragmatic advice to someone who needs it. To make your report credible, you have to play the role of a dispassionate expert, able to use the right terminology, cite the right sources, find and present hard evidence, and so on. But most of all, you have to design your report around a specific *intention* that shapes your role: to advise a reader about what he must *do* to solve his problem. That kind of research report is common in the world at large, but is much less common in the academic world than the following one.

2.2.3 I've Found an Answer to a Question Important to You

Although academic researchers sometimes offer advice to people like EPA directors, their most common role is that of the scholar, someone who answers questions so that a research community can simply understand its area of special interest better. Others might later use those answers to solve a practical problem—an arcane discovery about the distribution of prime numbers, for example, helped cryptologists design an unbreakable code. But the research itself aimed primarily at solving not a practical problem, but a *conceptual* one, one defined by incomplete knowledge or flawed understanding. Some researchers call this "pure" as opposed to "applied" research.

Teachers occasionally invent "real world" scenarios based on conceptual problems: a political science professor asks you to play the role of a senator's intern researching the effect of TV on children's intellectual growth. But more typically they expect you to imagine yourself as what you are learning to be—a researcher who can address an academic research community interested in a question that its members want to understand better. Your report on medieval Tibetan weaving, for example, might help explain some larger question not entirely understood, perhaps how medieval Tibetan art influenced modern Chinese art.

2.3 CREATING THE OTHER HALF OF THE RELATIONSHIP: THE READER'S ROLE

When you adopt one of those three roles, you create one half of the relationship between you and your readers. You create the other half when you write in a way that casts your readers in a complementary role, one giving them a specific reason to read your report. To do that, you have to imagine them as the kind of readers who expect you to do what you in fact intend to do. In creating those roles, you offer your readers a social contract: *I'll do my part if you do yours.* If you cast them in a role that they accept, but then you create one for yourself that doesn't match, you seem not to be upholding your end of the bargain. But if you offer them a role they are unwilling to adopt, you are likely to lose them entirely.

For example, suppose you are a researcher who is an expert on blimps and zeppelins. You have been invited to share your research with three different groups that have three different reasons for wanting to know what you know.

2.3.1 Entertain Me with Something Interesting I Didn't Know

Imagine that the first group that has invited you to speak is the local Zeppelin Club. Its members are fascinated with zeppelins, and though they know a lot about them, they are not experts, just ordinary folk who have made zeppelins their hobby. You decide to share some new facts you've dug up and to tell an entertaining tale or two. You read a letter from Great-Uncle Otto to your father describing a trip on a zeppelin in 1936, and you pass around some photographs and menus he saved.

In planning that report, you judge that not much is at stake in it other than a diverting hour of zeppelin lore. If so, you fulfill your side of the bargain when you tell them something about zeppelins that is new and interesting to *them*, even unsubstantiated folklore—and you don't bring along overheads, data tables, or footnotes to substantiate your sources. Your audience fulfills its role by listening with interest, maybe by sharing their own anecdotes. You don't expect them to challenge the authenticity of the letter or the menu or ask skeptical questions about how the photos and menus should change their wider understanding of the social history of zeppelins.

Some beginning researchers imagine their readers are like the Zeppelin Club—eager to hear any information new to them. While that sometimes works for experts who find the right audience (see the box below), it rarely works for students learning to do and report research. Your teachers assign you research projects to see not just what you can find, but what you can make of it.

2.3.2 Help Me Solve a Practical Problem

Now imagine that you have been invited to meet with the public relations department of Hotair.com. They suffer from low name recognition and want to use a blimp to get their logo before the public, flying it at sporting events, outdoor concerts, and other large gatherings. But they don't know whether that's a practical solution. So they have hired you as a consultant to tell them how much it will cost, how many days the weather is good enough to fly, and so on. For this group, you won't mention what Great-Uncle Otto had for dinner on his zeppelin flight in 1936. To succeed in this relationship, you must offer them a solution to their problem and only those facts that back it up.

That is the kind of situation you are likely to face if you have a job or internship, or if your teacher creates one of those scenarios for a "real world" writing assignment—you are an environmental scientist advising the state EPA about the polluted lake. Academic researchers do sometimes write on practical problems, but conceptual ones are far more common, even in applied disciplines like engineering. So pose a practical problem *only if* your teacher has created a specific scenario for one or you have checked with her first. (We'll discuss practical problems in more detail in the next chapter.)

2.3.3 Help Me Understand Something Better

Now imagine that your audience is the faculty of Zeppo University's Department of Lighter-than-Air Studies (with the same standing as, say, your departments of English or physics). They study the history of blimps and zeppelins, do research on their economics and aerodynamics, and participate in a worldwide conversation about their cultural history and social significance. They compete with one another in producing new knowledge and new lighter-than-air theories that they publish in lighter-than-air journals and books read by everyone in their field.

These scholars have invited you to talk about your specialty: transatlantic zeppelin flights in the late 1930s. They don't want you just to amuse them (though they will be happy if you do) or to help them *do* something (though they would be pleased to learn how to get consulting work with Hotair.com). What they most want is for you to tell them something they don't know about zeppelins, not just for its own sake, but so that they can better understand something new about them.

Because these lighter-than-air scholars are interested in the Truth about zeppelins, you know they will expect you to be objective, rigorously logical, faithful to the evidence, able to see every question from all sides. You also know that if you don't nail down the facts, they will hammer you during the question period afterward and during cocktails after that, not just to be contentious or even nasty (though some will be), but to get as close as they can to the Truth about zeppelins. If you offer something new, like Great-Uncle Otto's menus, they will want to know where and how you got them, and how those items contribute to their understanding of the topic. And to be sure they're the real thing, they will question you closely about how you know they are authentic.

More important, they will take an interest in those menus only if you can show them how they help answer a question important to their understanding of zeppelins, especially if you can convince them that they do not understand something about zeppelins as well as they thought. If you don't, they will ask you the most vexing question of all, *So what? Why should I care about your menus?*

So you begin your talk:

> As we all have been led to believe by a number of studies on the food service on transatlantic zeppelin flights in the 1930s (especially Schmidt 1986 and Kloepfer 1998), shellfish and other highly perishable items were never served because of fears regarding health. However, I have recently discovered a menu from the July 12, 1936, crossing of the *Hindenburg* indicating that oysters were served at dinner. . . .

That is the kind of conversation you join when you report research to a community of scholars, whether lighter-than-air or not. When you enter into this relationship with them, you must imagine them having this conversation with you in their minds: *Never mind whether your style is graceful (though I will admire your work more if it is); don't bother me with amusing anecdotes about your great-uncle Otto (though I like hearing them if they help me understand your ideas better); ignore whether what you know will make me rich (though I would be happy if it did). Just tell me something that I don't know so that I can better understand the topic of our common interest.*

Since your particular readers will be strongly inclined to adopt this third role, they will think you have fulfilled your side of the bargain only when you meet their expectations and answer their questions, only when you treat them as who they think they are. To be sure, the faculty over in chemistry or philosophy probably won't care much about your views on zeppelins, much less their meal service. *Who cares about the trivia they study over in the Lighter-than-Air Department?* But then you don't have much interest in their issues, either. You are concerned with your *particular* community of readers, with their particular interests and expectations. The trick is to get your research community to recognize and accept not only the role you've adopted for yourself, but the role you have cast for them—which means *you* first have to learn what kinds of roles they are willing to play. Several of the following chapters show you how to do that.

WHO CARES ABOUT *THAT*?

Academic researchers are regularly chided for their esoteric interests. That charge is usually unfair, but some researchers do seem to have a blinkered fascination with narrow objects of study. Williams once attended the dissertation defense of a Ph.D. candidate who had discovered reels and reels of silent film shot by European anthropologists in Africa and Asia in the early part of the twentieth century. No one had known that those films existed. These new data fascinated most of the examiners, film scholars who never questioned their worth. But when Williams asked, "But how does this discovery improve or even correct our understanding of movies then or now?" the candidate had no answer. She merely described again the specific content of the films, concluding, "And no one has ever seen this footage before." Williams asked his question in different ways but never got a better answer. The film scholars, on the other hand, were untroubled, because they, no doubt, were already thinking about how the footage might change their thinking about early film. Besides, they all love the movies. So sometimes new data alone are enough to interest the right readers. But if that candidate hopes to write a research report that gets anyone but a small group of specialists to care about her work, she will have to make an offer better than *Here's some new stuff*.

2.4 WRITING IN GROUPS

One of the best ways to see how the reader-writer relationship works in person is to share your writing in an organized group. A group is better at anticipating what your intended readers will expect and at predicting their responses. A group can also be more critical of its collective work than any individual can. Moreover, a group can bring more resources to bear on a project than someone working alone. So if your teacher does not set up writing groups, ask her to consider doing so. Or form a group on your own. At the least, recruit some friends to respond to your drafts as surrogate readers. (If you are trapped into working entirely alone, skip to 2.5, p. 30.)

2.4.1 Three Keys for Working Together Successfully

TALK A LOT. Create conditions that get you talking a lot. Set regular meeting times, share e-mail addresses and fax numbers: do what you can to ensure that you talk regularly. At your first meeting, work on telling your "elevator story"—how you would describe your project to a stranger in an elevator as it goes from the first to the twentieth floor. It should describe your question or problem, the kind of claim you expect to offer, and the kind of evidence that supports it. Practice your elevator story at every meeting (even with outside friends), until you can explain your project in a way that everyone thinks is clear and interesting. (You will find the next two chapters particularly useful for this.)

You should also talk about your intended readers. What do they know already, what is important to them, what do you want them to do with your report? Use our checklists to share ideas about readers (pp. 32–33), to ask questions systematically (pp. 45–49), and to reformulate them as a problem (pp. 49–52). The more your group talks together, the better you will write together. You will need to talk less if (like the three of us) you have already worked together and can anticipate how the others think. Yet in writing this book, we three still made scores of phone calls, exchanged hundreds of e-mail messages, and sat together a dozen times (sometimes traveling hundreds of miles to do so).

AGREE TO DISAGREE. Don't expect to agree 100 percent on every issue. You will differ over particulars, sometimes heatedly. In resolving those differences, your group can do its best thinking if everyone is explicit about what each believes and why. On the other hand, nothing impedes progress more than someone's insisting on *his* wording or on including only *her* data. If the first rule of writing in a group is to talk a lot, the second is to keep disagreements in perspective. When you disagree over minor issues with little impact on the whole, forget it.

ORGANIZE AND PLAN. The group should appoint a moderator, facilitator, coordinator, organizer—the job has different names and can either rotate or be permanently assigned. That person keeps track of the schedule, checks progress, moderates discussions, and when the group seems deadlocked, decides which way to go. Someone else should maintain a common outline that is updated regularly, first as a topic outline (p. 187), then as an outline of your argument (p. 139), and finally an outline of your points (p. 188). If your project needs lots of data, someone should maintain a schedule to gather them and a list of sources consulted and still to be consulted, annotated by how useful each source has been or might be. Everyone can stay up to speed if your updated outlines, notes on sources, and comments are put up on a website available to all.

2.4.2 Three Strategies for Working in Groups

Groups can organize their work in three ways, each of which has benefits and risks. Most groups combine these strategies.

DIVIDE AND DELEGATE. This strategy works best when tasks are parceled out to make best use of the special talents of each member. A group working on a survey, for example, might decide that two people are good at gathering data, two others at analyzing them and producing graphics, two more at drafting, and all will take a turn at revising. (Working on this revision, for example, one of us—of course, the youngest—was assigned responsibility for explaining how to use the Internet.) This strategy crucially depends on each member finishing tasks on time. If one fails, all fail.

A risky strategy is to assign whole sections of a document to different members to research, draft, and revise. That works only when the parts of a report are independent, but even then someone has to make the parts hang together, and that can be difficult if members have failed to consult along the way. And, if one fails to meet a deadline, all fail.

WORK SIDE BY SIDE. Some groups share all the work all the way. This works best with a small, tightly knit group working on a clearly defined project with ample time, like four engineering students devoting a semester to one design project. The disadvantage is that some people are uncomfortable talking about half-formed ideas before they work them out in writing. Others find it even more difficult to share drafts. To follow this strategy, members must be tolerant with one another. Too often, the most confident person ignores the feelings of others, dominates the process, and blocks progress.

TAKE TURNS. Once all the data have been gathered and an outline agreed on, some groups take turns drafting and revising, so that a text slowly evolves toward a final version. This strategy works when differences among members complement rather than contradict one another. For example, in a group working on a history of stories about the Alamo, one person might be interested in the clash of cultures, another in political consequences, and a third in the role of narrative in popular culture. After sharing what they find, they take turns working on the whole draft. One writer does a rough draft with enough structure so that others can see the shape of the argument. Each member in turn takes over the draft, adding ideas that seem important. The group must agree that the person working on the draft "owns" it while she has it and can change it however she wishes, so long as the changes reflect a common understanding of the main point the whole project supports.

This approach runs two risks. First, the final draft might zigzag from one interest to another. A group that works by turns must agree on a final goal and shape of the whole, and each member must respect the perspectives of the others. Second, you can lose track of who has revised what version of a draft. To avoid confusion, round-robin the drafts so that only one person is working on any one part of a draft at any one time and it is clear who gets the draft next.

Some groups use different strategies at different stages. In early planning, they work side by side until they form a general sense of their problem, then for data-gathering, they divide up the work, then take turns for revision. That's what we did in writing this book. Early on, we worked side by side until we had an outline, then assigned ourselves separate chapters. When the process stalled, we worked side by side again to revise our plan (that happened three times). Most often, though, each of us drafted individual chapters, then circulated drafts round-robin style. As a result, all of the chapters differ from the ones originally drafted, most quite a bit.

Whatever your strategy, the greatest risk is lack of coordination, so be clear who is supposed to do what and when. Then write it down and give everyone a copy. Working in groups is hard work, and it can be especially hard on the ego, but it can also reward those willing to listen to the sometimes harsh but usually helpful judgment of others.

2.5 MANAGING THE UNAVOIDABLE PROBLEM OF INEXPERIENCE

All researchers start as novices. We all face the uneasiness of trying to establish ourselves in a field whose basic rules we don't fully understand, much less all the subtle and unspoken rules that go into acting and writing like a member of our research community. Then, much to our surprise, we feel that novice anxiety again when we begin a new project on a topic that we don't know much about. We three authors have felt those anxieties, not just starting out, but long after our hair had grayed. No one can avoid feeling overwhelmed and anxious at times, but there are some things you can do about it:

- First, be aware that there are uncertainties and anxieties that you cannot avoid. You can learn something about them from a first quick reading of this book. Get over those you can, but don't hold it against yourself when you feel anxious. It is not a sign of incompetence but of inexperience.

- Second, get control over your topic by writing about it along the way. Don't just retype or photocopy sources: write summaries, critiques, questions. The more you write as you go, no matter how sketchily, the more confidently you will face that intimidating first draft.

- Third, understand the whole process by breaking it into manageable steps, but be aware that those steps are mutually supportive. Once you find a topic and formulate a good question, you'll draft and revise more effectively. Conversely, if you anticipate how you will draft and revise, you can more effectively find a problem now.

- Fourth, count on your teacher to understand your struggles. Good teachers want you to succeed, and you can expect their help. (If they don't help, look for other mentors whom you might consult.)

Finally, set realistic goals. You do something significant when you wind up your project feeling that you have changed what *you* think and that your readers think you did it soundly, even if they don't agree. Most important, recognize the struggle for what it is—a learning experience. To overcome the problems that all beginners face, do what successful researchers do, especially when discouraged: press on, confident that it will turn out OK. Perhaps only "OK—considering." But perhaps even better than OK.

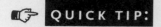 *A Checklist for Understanding*
Your Readers

Think about your readers from the start, knowing that you'll understand them better as you work through your project. Answer these questions early on, then revisit them when you start planning and again when revising.

1. Who will read my report?

   - Professionals?

   - General readers who are well informed?

   - General readers who know little about the topic?

2. Do they expect me to do what I intend to do? Should I

   - entertain them?

   - provide new factual knowledge?

   - help them understand something better?

   - help them do something to solve a practical problem in the world?

3. How much can I expect them to know?

   - What do they know about my topic?

   - What special interest do they have in it?

   - What are they likely to expect me to discuss?

   - Is the problem one that they already recognize?

   - Is it one that they have but haven't yet recognized?

   - Is the problem not theirs, but only mine?

- Will they automatically take the problem seriously, or must I labor to convince them that it matters?

4. How will readers respond to the solution/answer in my main claim?

   - Will it contradict what they already believe? How?

   - Will they know some standard arguments against my solution?

   - Will they want to see the steps that led me to the solution?

   - Do they expect my report to follow a standard format? If so, what is it?

PART II

Asking
Questions,
Finding
Answers

Prologue

If you've skimmed this book once, you're ready to begin your project. If you already have a question and know how to answer it, review the next two chapters; then before you start drafting, read the remaining chapters carefully. If, on the other hand, you are starting from scratch, with no clear direction, not even an assigned topic, you may feel bewildered. But you can manage if you have a plan to guide you through your project, one step at a time.

Unfortunately, no plan can lead you straight to that finished report. Early on you may have to spend time reading randomly just to discover what interests you. You may wander up blind alleys or lose yourself in heaps of data. But if you have a plan, it can guide you through that confusion (or even help you avoid it).

Your first four steps in planning are these:

1. Find a topic specific enough to let you master a reasonable amount of information on it: not, for example, *the history of scientific writing*, but *essays in the Proceedings of the Royal Society (1675–1750) as precursors to the modern scientific article;* not *doctors in seventeenth-century drama*, but *Molière's mockery of doctors in his early plays.*

2. Ask questions about that topic until you find some that catch your interest. For example, *How did early Royal Society*

authors guarantee the reliability of their evidence? Or, *How do the differences between their procedures and modern ones reflect differences in the social structure of science?* Or, *Why were doctors objects of Molière's mockery?*

3. Determine what *kind* of evidence that your readers will expect in support of your answer. For example, will they accept data from secondary sources, or will they expect you to consult primary sources as well? Will they expect quantitative data or quotations from authorities?

4. Determine whether you can find sources that have those data.

Once you see in the data that you find at least a plausible answer to your question, you'll be ready to start shaping your materials into an argument (the subject of part III), then to draft and revise it (the subjects of part IV).

Expect to do lots of writing along the way. Much of it will be routine note-taking, but you should also spend time writing to understand: make preliminary outlines; disagree with what you have read; draw diagrams to connect disparate facts; summarize sources, positions, and schools; record even random thoughts. You never know what will pay off. You probably won't include much of this preliminary writing in your final draft; you may even discard it all and start over. But if you *write as you go,* you'll encourage your own best critical thinking, understand your sources better, and draft more effectively when that time comes.

You will discover, however, that you cannot move through those four steps in the neat order we presented them. You'll probably think of a tentative answer and outline a supporting argument before you have all the evidence you need. And when you think you have an argument worth making, you'll probably decide that you need more and maybe different evidence from new sources. You may even modify your topic. Doing research is not like strolling along a well-marked path to a familiar destination; it's more like struggling through overgrown woods, searching for something you won't know until you find it. But no matter how indirect your path, you can feel confident that you are steadily getting closer to an answer if you manage each step of the way to anticipate the predictable problems.

WHAT ARE YOUR DATA?

No matter their field, researchers collect information to use as evidence in support of their claims. But researchers in different fields call that information by different names. Here, we use the term *data*. By *data* we mean more than the numbers that natural and social scientists collect. We mean anything you find "out there" that might support your answer to a question or solution to a problem. The term is rarely used by researchers in the humanities, but they, too, gather data in the form of quotations, historical facts, and so on. Data are inert, however, until you use them as *evidence* to support a *claim*. If you have not collected more data than you can use, you haven't found enough. (Incidentally, remember that *data* is plural; a single bit of data is a *datum*).

CHAPTER THREE

From Topics to Questions

In this chapter we discuss how to explore your interests to find a topic, narrow it to a manageable scope, question it to find the makings of a problem, then turn it into a problem that guides your research. If you are an experienced researcher or already know what topics you want to pursue and why, you might skip to chapter 4. But if you are starting your first project, you will find this chapter useful.

If you are free to research any topic that interests you, that freedom can be frustrating—so many choices, so little time. At some point, you have to settle on a topic, but beyond a topic, you also have to find a reason beyond your assignment to devote weeks or months pursuing it and writing up what you find, then to ask readers to spend their time reading your report.

As we've said, your readers expect you to do more than just mound up and report data; they expect you to report it in a way that continues the ongoing conversation between writers and readers that creates a *community* of researchers. To do that, you must select from all the data you find just those data that support an answer to a question that solves a problem your readers think needs solving. In all research communities, some problems are already "in the air," widely debated and deeply researched, such as whether personality traits like shyness or an attraction to risk are genetically inherited or learned. But other questions may intrigue only the researcher: *Why do cats rub their faces against us? Why do the big nuts end up at the top of the can?* That's how a lot of research begins—not with a "big" question known to everyone in a field, but with a mental itch that only one researcher feels the need to scratch.

If you have such an itch, good. But as we've said (and will say again), at some point, you have to decide whether the answer to your private question is also significant to others: to a teacher, colleagues, other researchers, or even to a public whose lives your research could change. At that point, you aim not just to answer a question, but to pose and solve a *problem* that others also think is worth solving.

Now that word *problem* is itself a problem: commonly, a problem means trouble, but among researchers it has a meaning so special that we devote all of the next chapter to it. It raises issues that few beginning researchers are able to resolve entirely and that can vex even advanced ones. But before you can address a research problem, you have to find a topic that might lead to one. We'll start there, with finding a topic.

3.1 FROM AN INTEREST TO A TOPIC

Most of us have more than enough interests to pursue, but beginners often find it hard to locate among theirs a topic focused enough to support a research project. A research topic is an interest defined narrowly enough for you to imagine becoming a local expert on it. That doesn't mean that you already know a lot about it or that you will have to learn more about it than your professor has. You just want to know more than you do now.

If your assignment leaves you free to explore any topic within reason, we can offer only a cliché: Start with what interests you most deeply. Nothing contributes to the quality of your work more than your commitment to it. Start by listing two or three interests that you'd like to explore. If you are undertaking a research project in a course in a specific field, skim a recent textbook, talk to other students, or consult your teacher. You might try to identify an interest based on work you are doing or will do in a different course.

If you are still stuck, you can find help either on the Internet or in your library. The Internet may seem the easier way, but it's more likely to lead you astray, especially if you are new to research. Start with the standard guides:

- For a project in a general writing course, start in the library. Look at the headings in a general bibliography such as the *Reader's Guide to Periodical Literature*. If you already have a general focus, use more specialized guides such as the *American Humanities Index* or the *Chicano Index*. (We discuss using these resources in chapter 5 and list many of them on pp. 298–315.)

Scan headings for topics that catch your interest. They will provide not only possible topics, but up-to-date references on them. If you already have an idea for a topic, you can check out the Internet, but if you have no idea what you are looking for, what you find there may overwhelm you. Some indexes are available online, but most don't let you skim only subject headings.

- For a first research project in a particular field, skim headings in specialized indexes, such as the *Philosopher's Index*, the *Psychological Abstracts*, or *Women's Studies Abstracts*.

Once you identify a general area of interest, use the Internet to find out more about it and to help you narrow your topic. (If you are really stuck, see the Quick Tip at the end of this chapter.)

- If you are doing an advanced research project, you might look first for what resources are easily available *before* you settle on a topic.

If you pick a topic and then discover that sources are hard to find, you may have to start over. If you *first* identify resources available in your library or on the Internet, you can plan your research more efficiently, because you will know where to start.

At first, you may not know enough about a general interest like *the use of masks in religious and social contexts* to turn it into a focused topic. If so, you have to do some reading to know what to think about it. Don't read randomly: start with entries in a general encyclopedia, then look at entries in a specialized encyclopedia or dictionary, then browse through journals and websites until you have a grip on the general shape of your topic. Only then will you be able to move on to these next steps.

3.2 FROM A BROAD TOPIC TO A FOCUSED ONE

At this point, you risk settling on a topic so broad that it could be a subheading in an encyclopedia: *Space flight, history of; Shakespeare, problem plays; Natural kinds, doctrine of.* A topic is usually too broad if you can state it in four or five words:

Free will in *War and Peace*	The history of commercial aviation

With a topic so broad, you may be intimidated by the idea of finding, much less reading, even a fraction of the sources available. So you have to narrow it, like this:

Free will in *War and Peace* ➞	The conflict of free will and historical inevitability in Tolstoy's description of three battles in *War and Peace*
The history of commercial aviation ➞	The crucial contribution of the military in the development of the DC-3 in the early years of commercial aviation

We narrowed those topics by adding words and phrases, but of a special kind: *conflict, description, contribution,* and *development.* Those nouns are derived from verbs expressing actions or relationships: *to conflict, to describe, to contribute,* and *to develop.* Without such words, your topic is a static thing—*free will in War and Peace, the history of commercial aviation.* But when you use nouns derived from verbs, you move your topic a step closer to a *claim* that your readers might find significant.

Note what happens when these topics become statements. Topics (1a) and (2a) change almost not at all:

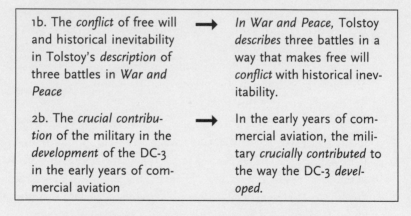

TOPIC	CLAIM
1a. Free will and historical inevitability in Tolstoy's *War and Peace*? ⟶	There is free will and historical inevitability in Tolstoy's *War and Peace*.
2a. The history of commercial aviation ⟶	Commercial aviation has a history.

Topics (1b) and (2b), on the other hand, are closer to claims that a reader might find interesting:

1b. The *conflict* of free will and historical inevitability in Tolstoy's *description* of three battles in *War and Peace* ⟶	*In War and Peace*, Tolstoy *describes* three battles in a way that makes free will *conflict* with historical inevitability.
2b. The *crucial contribution* of the military in the *development* of the DC-3 in the early years of commercial aviation ⟶	In the early years of commercial aviation, the military *crucially contributed* to the way the DC-3 *developed*.

Such claims will at first seem weak, but you will develop them into more specific ones as you develop your project.

A more specific topic also helps you see gaps, puzzles, and inconsistencies that you can ask about when you turn your *topic* into a research *question* (more about that in a moment). A specific topic can also serve as your working title, a short answer when someone asks you what you are working on.

Caution: Don't narrow your topic so much that you can't find enough data on it:

TOO MANY DATA AVAILABLE	TOO FEW DATA AVAILABLE
The history of commercial aviation	The decision to lengthen the wingtips on the DC-3 prototype because the military wanted to use the DC-3 as a cargo carrier

3.3 FROM A FOCUSED TOPIC TO QUESTIONS

In taking this next step, researchers often make a beginner's mistake: they rush from a topic to a data dump. Once they hit on a topic that feels promising, something like *the political origins and uses of legends about the Battle of the Alamo,* they go straight to searching out sources—different versions of the story in books and films, Mexican and American, nineteenth century and twentieth. They accumulate a mound of summaries of the stories, descriptions of their differences and similarities, ways in which they conflict with what modern historians think happened. They write all that up and conclude, "Thus we see many interesting differences and similarities between . . ."

Most high school teachers would give such a report a passing grade, because it shows that the student can focus on a topic, find data on it, and assemble those data into a report—no small achievement for a first project. But in any advanced course, including a first-year writing course in college, such a report falls short because it offers only random bits of information. If the writer asks no *question* worth pondering, he can offer no focused answer worth reading. Readers of research reports don't want just information; they want the answer to a question worth asking. To be sure, those fascinated by a topic often feel that *any* information about it is worth reading for its own sake: collectors of Japanese coins or Elvis Presley movie posters will read anything about them. Serious researchers, however, do not report data for their own sake, but to support the answer to a question that they (and they hope their readers) think is worth asking.

The best way to find out what you do not know about a topic is to barrage it with questions. First ask the predictable ones of your field. For example, a historian's first questions about the Alamo stories would concern their sources, development, and accuracy. Also ask the standard journalistic questions *who, what, when,* and *where,* but focus on *how* and *why.* Finally, you can systematically ask four kinds of analytical questions, about the composition, history, categorization, and values of your topic. Record the questions, but don't stop for answers. (And don't worry about fitting the questions into the right categories; use the categories only to stimulate you to ask them and to organize their answers.)

3.3.1 Identify the Parts and How They Interrelate

- What are the parts of your topic, and how do they relate to one another?

  In stories about the Alamo, what are the themes, the plot structure, the main characters? How do the characters relate to the plot, the plot to the actual battle, the battle to the characters, the characters to one another?

- How is your topic part of a larger system?

  How have politicians used the story? What role does it have in Mexican history? What role does it have in U.S. history? Who told the stories? Who listened? How does their nationality affect the story?

3.3.2 Trace Its Own History and Its Role in a Larger History

- How and why has your topic changed through time, as something with its own history?

  How have the stories developed? How have different stories developed differently? How have audiences changed? How have the storytellers changed? How have their motives to tell the stories changed?

- How and why is your topic an episode in a larger history?

    How do the stories fit into a historical sequence of events? What caused them to change? How did they affect national identity in the United States? In Mexico? Why have they endured so long?

3.3.3 Identify Its Characteristics and the Categories that Include It

- What kind of thing is your topic? What is its range of variation? How are instances of it similar to and different from one another?

    What is the most typical story? How do others differ? Which is most different? How do the written and oral stories differ from the movie versions? How are Mexican stories different from those told in the States?

- To what larger categories can your topic be assigned? How does that help us understand it?

    What other stories in U.S. history are like the story of the Battle of the Alamo? In Mexican history? How do the stories compare to other mythic battle stories? What other societies produce similar stories?

3.3.4 Determine Its Value

- What values does your topic reflect? What values does it support? Contradict?

    What moral lesson does the story teach, if any? Whose purposes does each story serve? Who is praised? Who blamed? Why?

- How good or bad is your topic? Is it useful?

    Are some stories better than others? More sophisticated than others? What version is the best one? The worst one? Which parts are most accurate? Which least?

3.3.5 Evaluate Your Questions

When you run out of questions (or think, *Enough!*), it's time to evaluate them. First, set aside questions whose answers you could look up in a reference work. Questions that ask *who, what, when,* or *where* are important, but they may ask only about matters of settled fact (though not always). Questions that ask *how* and *why* are more likely to invite deeper research and lead to more interesting answers.

Next, try to combine smaller questions into larger, more significant ones. For example, several Alamo questions revolve around the issue of the interests of the storytellers and their effects on the stories:

> How have politicians used the story? What role does it have in U.S. history? How have the storytellers changed? How have their motives to tell the stories changed? How did the stories affect national identity in the United States? How do the stories compare to other mythic battle stories? Is its moral lesson worth teaching? Whose purposes does each story serve?

Many of these can be combined into a larger, more significant question:

> How and why have tellers of the Alamo story given a mythic quality to the event?

Once you settle on a question or two, you have a guide to doing your research more systematically. A question narrows your search to only those data you need for its answer. And once you have an answer you think you can support, you know it's time to stop hunting. But when you have only a topic, the data you can find on it are, literally, endless; worse, you will never know when you have enough.

Through all this, though, the most important goal is to find questions that challenge you or, better, arouse your intense curiosity. Of course, you can't be sure where any particular question will lead, but this kind of questioning can send you in directions you never imagined, opening you up to new interests, new worlds of research. Finding good questions is an essential step in any project that goes beyond fact-grubbing. With one or two in mind, you are ready for the next steps.

3.4 FROM A MERELY INTERESTING QUESTION TO ITS WIDER SIGNIFICANCE

Even if you are an experienced researcher, you might not be able to take this next step until you are well into your project. If you are a beginner, you may feel that this step is still deeply frustrating even when you've finished it. Nevertheless, once you have a question that grabs your interest, you must pose a tougher question: *Why should this question also grab my readers? What makes it worth asking?*

Start by asking, *So what?* At first, ask it for yourself:

> So what if I don't know or understand how snow geese know where to go in the winter, or how fifteenth-century violin players tuned their instruments, or why the Alamo story has become myth? So what if I can't answer those questions?

Eventually, you will have to answer this question not just for yourself but for your readers. Finding its answer vexes all researchers, beginners and experienced alike, because it's so hard to predict what will really interest readers. Instead of trying to answer instantly, though, you can work toward an answer in three steps.

3.4.1 Step 1: Name Your Topic

If you are just beginning a project, with only a topic and maybe the glimmerings of a few good questions, describe your topic in a sentence as specific as you can make it (glance back at pp. 43–45):

> I am trying to learn about (working on, studying) _____.

Fill in the blank with your topic. Be sure to use some of those nouns based on verbs or adjectives:

I am studying *diagnostic processes* in the *repair* of cooling systems.

I am working on Lincoln's *beliefs* about *predestination* in his early speeches.

3.4.2 Step 2: Add a Question

As soon as you can, add to that sentence an indirect question that specifies something that you do not know or understand about your topic but want to:

1. *I am studying X*
   2. *because I want to find out* who/what/when/where/whether/ why/how _____.

1. *I am studying* diagnostic processes in the repair of cooling systems
   2. *because I am trying to find out how* expert repairers diagnose failures.

1. *I am working on* Lincoln's beliefs about predestination in his early speeches
   2. *because I want to find out how* his belief in destiny influenced his understanding of the causes of the Civil War.

When you add that *because-I-want-to-find-out-how/why* clause, you state why you are pursuing your topic: to answer a question important to you.

If you are doing one of your first research projects and you get this far, congratulate yourself, because you have framed your project in a way that moves it beyond the kind of aimless collection and reporting of data that afflicts too much research. But now go one step more, if you can.

3.4.3 Step 3: Motivate Your Question

This step is a hard one, but it lets you know whether your question is not just interesting to you but possibly significant to others. To do that, add another indirect question, a bigger and more general one that explains why you are asking your first question.

Introduce this second implied question with *in order to help my reader understand how, why,* or *whether:*

1. *I am studying* diagnostic processes in the repair of cooling systems
   2. *because I am trying to find out how* expert repairers analyze failures,
      3. *in order to help my reader understand how* to design a computerized system that can diagnose and prevent failures.

1. *I am working on* Lincoln's beliefs about predestination in his early speeches
   2. *because I want to find out how* his belief in destiny and God's will influenced his understanding of the causes of the Civil War,
      3. *in order to help my reader understand how* his religious beliefs may have influenced his military decisions.

It's your answer to the third step that will give you a claim on your readers' interest. If that larger question touches on issues important to your field, even indirectly, then you have reason to think that your readers should care about its answer, and so care about your answer to the smaller, prior question you raise in step 2.

A few researchers can flesh out this whole pattern even before they start gathering data, because they are working on a well-known question, some widely investigated problem that others in their field are already interested in. In fact, advanced researchers often begin their research with questions that others have asked before but not answered thoroughly, or maybe even correctly. But many researchers, including at times the three of us, find that they can't flesh out these steps until they're nearly finished. And too many write up their research results without having thought through these steps at all.

At the beginning of your project, you may not be able to get past the first step of naming your topic. But regularly test your progress by asking a roommate, relative, or friend to *force* you to question your topic and to flesh out those three steps. Even if you can't take them all confidently, you'll know where you are and where you still have to go.

To summarize: Your aim is to explain

1. what you are writing about—your topic: *I am studying* . . .

2. what you don't know about it—your question: *because I want to find out* . . .

3. why you want your reader to know about it—your rationale: *in order to help my reader understand better* . . .

If you are just beginning serious research, don't be discouraged if you never get past that second step. As long as your question is interesting to *you,* plow ahead. Your teacher should be satisfied, because you have changed the terms of your project from simply gathering data to asking and answering a question.

If you are a graduate student doing advanced research, however, you *must* take that last step, because answering that last question will help you create the relationship you are working to establish with the rest of your research community. It's your ticket into the conversation.

In the following chapters, we will return to those three steps and their implied questions, because as you'll see, they are crucial not just for finding good specific questions that you want to answer, but for finding and then expressing the problem that you want your readers to recognize and value.

If you have experience in your field but are stuck for a topic, you can find one with some quick research. Read recent articles and review essays and, if they are available, recent dissertations. Look closely at the conclusions: they often suggest further lines of research. You can also browse the archives of an Internet discussion list in your field: look for points of current controversy.

But if you are a beginner and your teacher has not suggested specific topics, start with our suggestions about skimming bibliographical guides (pp. 298–315). If you still draw a blank, try these steps.

FOR GENERAL TOPICS

1. What special interest do you have—sailing, chess, finches, old comic books? The less common, the better. Investigate something about it you don't know: its origins, its technology, how it is practiced in another culture, and so on.

2. Where would you like to go? Surf the Internet, finding out all you can about it. What particular aspect surprises you or makes you want to know more?

3. Wander through a museum with exhibitions that appeal to you—artworks, dinosaurs, automobiles. If you can't get there in person, browse a "virtual museum" on the Internet. Stop when something catches your interest. What more do you want to know about it?

4. Wander through a shopping mall or store, asking yourself, *How do they make that?* or, *I wonder who thought up that product?*

5. Leaf through a Sunday newspaper, especially its features sections, until something catches your eye. Skim reviews of books or movies, in newspapers or on the Internet.

6. Browse a large magazine rack. Look for trade magazines or those that cater to specialized interests. Investigate whatever catches your interest.

7. If you can use an Internet newsreader, look through the list of "alt" newsgroups until you find one that sounds interesting. Read the posts, looking for something that surprises you or that you disagree with.

8. Tune into talk radio or interview programs on TV until you hear a claim you disagree with. Or find something to disagree with on the websites connected with well-known talk shows. See whether you can make a real case to refute it, instead of just shouting back.

9. Use an Internet search engine to find websites about something people collect. (Narrow the search to exclude dot-com sites.) You'll get hundreds of hits, but look only at the ones that surprise you.

10. Is there a common belief that you suspect is much too simplistic, or just plain wrong? Or a common practice that you detest? Don't just pronounce the belief or practice wrong, but instead probe for something you can show about it that might lead others to reconsider.

FOR TOPICS FOCUSED ON A PARTICULAR FIELD

1. Browse through a textbook of a course that is one level beyond yours or a course that you know you will have to take some time in the future. Look especially hard at the study questions.

2. Attend a lecture for an advanced class in your field and listen for something you disagree with, don't understand, or want to know more about.

3. Ask your instructor about the most contested issue in your field.

4. Find an Internet discussion list in your field. Browse its archives, looking for matters of controversy or uncertainty.

5. Surf the websites of departments at major universities, including class websites. Also check sites of museums, national associations, and government agencies, if they seem relevant.

CHAPTER FOUR

From Questions to Problems

In this chapter we explain how to frame your project as a problem that readers want to see solved, an essential step for advanced researchers. If you are attempting your first research project, this chapter may prove difficult. (You can find more help on problems in our discussion of introductions in chapter 14.) If you feel lost, you can skip to chapter 5, but we hope that you will stay with it. You'll learn important steps you can take now, and will certainly need in the future.

In the last chapter, we described how to find a topic in your interests, how to narrow it, then to question it. We suggested that you identify the significance of your questions by fleshing out this three-step formula:

1. **Topic:** I am studying _____
   2. **Question:** because I want to find out what/why/how _____,
      3. **Significance:** in order to help my reader understand _____.

These steps describe not only the development of your project, but your own as a researcher.

When you move from step 1 to 2, you stop being a mere data collector, because you are now motivated not by aimless curiosity (by no means a useless impulse), but by a desire to understand something better. That second step also helps you develop an increasingly sophisticated relationship with your readers. When you move from step 2 to 3, you focus your project on the significance of that understanding, at least for yourself. But you can join a *community* of researchers only when you can see that significance from your readers' point of view. With that last step, you change your intention from merely discovering and understanding something for yourself to *showing* and *explaining* something to others, a move that makes a stronger claim on readers and so creates a stronger relationship with them.

4.1 PROBLEMS, PROBLEMS, PROBLEMS

That third step is hard for everyone, even experienced researchers. Too many write as if they do their job by answering a question that happens to interest them. They fail to understand that their answer must also solve a *problem* that is significant to their community of readers. But researchers often cannot start their project knowing exactly what problem they will finally solve. Many start with just a hunch, a puzzle, something they want to know more about; it's not until they are well into their research, sometimes even their drafting, that they finally figure out what problem they have solved. So don't feel uneasy if early in your project you do not yet know exactly the significance of your question. But you can begin planning your research knowing (or at least hoping) that a good one is out there somewhere.

To understand how to find that good question and then communicate its significance, though, you first have to know what a research problem really is.

4.1.1 Practical Problems and Research Problems

Everyday research usually begins not with dreaming up a topic but with solving a practical problem that has just landed on you, a problem that, left unresolved, means trouble. When the solution is not obvious, you ask questions whose answers you hope will help you solve it. But to answer them, you must pose and solve a problem of another kind, a *research* problem defined by what you do not know or understand, but feel you must before you can solve your practical problem.

This process of addressing practical problems is familiar. It typically looks like this:

PRACTICAL PROBLEM: My brakes have started screeching.
RESEARCH QUESTION: Where can I get them fixed right away?

RESEARCH PROBLEM: Find the Yellow Pages and look up closest brake shop.
RESEARCH ANSWER: The Car Shoppe, 1401 East 55th Street.
APPLICATION: Call to see when they can fix them.

It's a pattern common in every part of our lives:

The National Rifle Association presses me to oppose gun control. *Will I lose my election if I don't?* Take a poll. *A majority of my constituents support gun control.* Now decide whether to reject the NRA's request.

Costs are up at the Omaha plant. *What has changed?* Compare personnel before and after. *More turnover now.* If we improve training and morale, our workers stay with us. OK, let's see if we can afford to do it.

Problems like those rarely require us to write up a solution. We write only when we have to convince others that we've found and solved a problem important to *them:*

To manager of Omaha plant: Costs are up in Omaha because we have too much turnover. Employees see no future in their jobs and are quitting after a few months. To retain workers, we must upgrade their skills so they will want to stay. Here is a plan . . .

But before anyone could solve the *practical* problem of rising costs, someone had to pose and solve a *research* problem defined by not knowing why they were rising. Only then can they decide what to *do* about it.

Graphically, the relationship between practical and research problems looks like this:

4.1.2 Distinguishing Practical Problems and Research Problems

Though solving a practical problem usually requires that we solve a research problem as well, it is crucial to distinguish between them, because we solve and write about them in different ways.

- A *practical* problem is caused by some condition in the world, from e-mail spam to terrorism, that makes us unhappy because it costs us time, money, respect, security, pain, even our lives. You solve a practical problem by *doing* something that changes the world by eliminating the causes that lead to its costs, or by encouraging others to do so.

- A *research* problem is motivated not by palpable unhappiness, but by incomplete knowledge or flawed understanding. You solve it not by changing the world but by understanding it better.

Though a research problem is often motivated by a practical problem, you don't solve the practical problem just by solving the research one. The manager of the Omaha plant might know the answer to the research question *Why are costs rising?* but still struggle to solve the practical problem *How do we improve training?*

The term *problem* thus has a special meaning in the world of research, one that sometimes confuses beginners. In our everyday world, a practical problem is something we try to avoid. But in the academic world, a research problem is something we eagerly seek out, even inventing one, if we have to. Indeed, a researcher without a good research problem has a bad practical one, because with no research problem to work on, she has nothing to do.

There is a second reason inexperienced researchers sometimes struggle with this notion of a research problem. Experienced researchers often talk about their research problems in shorthand. When asked what they are working on, they respond with what sounds like one of those general topics we warned you about in the last chapter: *adult measles, early Aztec pots, the mating calls of Wyoming elk.*

As a result, some beginners think that having a topic to read about is the same as having a problem to solve. But when they do, they have a big practical problem, because without a research problem, they lack the focus set by the need to answer a particular question. So they gather data aimlessly and endlessly, with no way of knowing when they have enough. Then they struggle to decide what to include in their report and what not, finally throwing in everything they have, just to be on the safe side. So it's not surprising that they feel frustrated when a reader says, *I don't see the point; this is just a data dump.* To avoid that judgment, you need a problem to focus your attention on those particular data that will help you solve your problem. That means first understanding how problems work.

4.2 THE COMMON STRUCTURE OF PROBLEMS

Practical problems and research problems have the same basic structure. Both have two parts:

1. a situation or *condition,* and

2. the undesirable consequences of that condition, *costs* you don't want to pay.

What distinguishes them is the nature of those conditions and costs.

4.2.1 The Nature of Practical Problems

A flat tire is a typical practical problem, because it is (1) a condition in the world (the flat) that (2) exacts on you a tangible cost that you don't want to pay (not getting to work on time or missing a dinner date). But suppose you were bullied into the date and would rather be anywhere else. In that case, the flat has no significant cost; in fact, since it turns out to be a benefit, it is not a problem at all, but a solution. No cost, no problem.

For a practical, tangible problem, the condition can be literally anything, even winning the lottery. Suppose you win a million dollars but owe a loan shark two million, and your name gets in the paper. He finds you, takes your million, and breaks your leg. Winning a million turns out to be a Big Problem.

To pose a practical problem, you must be able to describe both its parts:

- its **condition**

  I missed the bus.

  The hole in the ozone layer is growing.

- the **costs** of that condition that make you (or someone) unhappy

  I will be late for work and may lose my job.

  Many will die from skin cancer.

But now a crucial caution: *Your readers* will judge the significance of a problem not by its cost to *you*, but by its cost to *them*. So you must try to frame your problem from their point of view. To do that, imagine that when you pose the condition of your problem, your reader responds, *So what?* For example,

The hole in the ozone layer grew last year.
So what?

You answer with the cost of the problem:

A bigger hole in the ozone means more ultraviolet light hitting the earth.

Suppose the other person again says, *So what?*, and you respond with a further cost:

Too much ultraviolet light can give people skin cancer.

If, however improbably, he again asks, *So what?*, you have failed to convince him that the problem is not just yours, but his as well. We acknowledge that a problem exists only when we stop saying, *So what?*, and instead say, *Oh no! What do we do about that?*

Practical problems like cancer are easy to grasp because they always have palpable consequences. In the academic world, however, you probably will work more with research problems, which are harder to grasp because both their conditions and costs are always abstract.

4.2.2 The Nature of Research Problems

Practical and research problems have the same structure, but their conditions and costs differ in important ways:

- The condition of a practical problem can be any state of affairs whose cost makes you (or someone) unhappy. The condition of a research problem, on the other hand, is *always* some version of *not knowing* or *not understanding* something.

You can identify conditions by working though the formula in chapter 3. The second step states what you do not know or understand:

I am studying stories of the Alamo *because I want to understand why voters responded to them in ways that served the interests of local Texas politicians.*

That's why we emphasized the value of questions. They force you to consider what you don't know or understand but want to.

- The cost of a practical problem is unhappiness. The consequence of a research problem, on the other hand, *is something else* we or, more important, our readers don't know or understand, but is *more* significant, *more* consequential than the ignorance or misunderstanding named by the condition. This, too, we can express as a question.

You identify these consequences in step 3 of our formula:

I am studying stories of the Alamo because I want to understand why voters responded to them in ways that served the interests of local Texas politicians, *in order to help readers understand how regional self-images influence national politics.*

All this may sound confusing, but it's simpler than it seems. When you move from questions to problems, you only translate that formula for working out the significance of a question *to you* into a way to find its significance *to your readers.*

It works like this: The first part of a research problem is something you don't know but want to. You can phrase that as a direct question:

How many stars are in the sky?

How have romantic movies changed in the last fifty years?

Now imagine someone asking, *So what if you can't answer that question?* What do you say? You answer by stating *something else* you don't know until you answer the first question, something that the other person should also want to know. For example,

If we can't answer the question of how romantic movies have changed in the last fifty years,*condition/first question* then we can't answer a more important question: How have our cultural depictions of romantic love changed?*consequence/larger, more important question*

If you think that finding an answer to that second question is important, you've stated a cost that makes your research problem worth pursuing, and if your reader thinks so too, you're in business.

But what if your potential readers might again ask, *So what?*

So what if I don't know whether our cultural depictions of romantic love have changed?

You will just have to pose a yet larger question whose answer depends on answering the previous ones, an answer that should be even more significant to your readers:

If we can't answer the question of how our cultural depictions of romantic love have changed in the last fifty years,*second question* then we can't answer a more important one yet: How is our culture shaping the expectations of young men and women concerning marriage and families?*consequence/larger, more important question*

If you imagine that reader again asking, *So what?*, you might be tempted to think, *Wrong audience*. But if that's the audience you're stuck with, you will have to try again.

To those outside an academic field looking in, researchers sometimes seem to pose a question so narrowly that outsiders think it is ridiculously trivial: *So what if we don't know how hopscotch originated?* Yet for those few who care about the way folk games influence the social development of children, the cost of not knowing justifies the research. *What do you mean? If we can discover how children's folk games originate, we can learn something about how they socialize themselves. . . .*

4.2.3 Distinguishing "Pure" and "Applied" Research

When the solution to a research problem has no apparent application to any practical problem in the world, but only to the scholarly interests of a community of researchers, we call the research *pure*. When the solution to a research problem does have practical consequences, we call the research *applied*.

You can tell whether a research problem is pure or applied by looking at the last of the three steps in defining your project. Does it refer to knowing or doing?

1. **Topic:** I am studying the density of light and other electromagnetic radiation in a small section of the universe
    2. **Question:** because I want to find out how many stars are in the sky,
        3. **Significance:** in order to help readers understand whether the universe will expand forever or contract into a new big bang.

That is a pure research problem because step 3 refers only to understanding.

In an applied research problem, the second step also refers to knowing, but that third step refers to *doing:*

1. **Topic:** I am studying the difference between readings from the Hubble telescope in orbit above the atmosphere and readings for the same stars from earthbound telescopes

2. **Question:** because I want to find out how much the atmosphere distorts measurements of light and other electromagnetic radiation,

   3. **Practical Significance:** so that *astronomers can use data from earthbound telescopes to measure more accurately the density of electromagnetic radiation.*

That is an applied problem because astronomers can *do* what they need to—measure light more accurately—only when they *know* how much atmospheric distortion to account for.

4.2.4 Connecting a Research Problem to Practical Consequences

Some less experienced researchers are uncomfortable with pure research because its costs—merely not knowing something—are so abstract. Since they are not yet part of a community that cares about the answers to their questions, they feel that their findings aren't good for much. So they try to cobble a practical cost onto their conceptual research question to make it seem more significant:

1. **Topic:** I am studying the differences among various nineteenth-century versions of the story of the Alamo

2. **Conceptual Question:** because I want to find out how politicians used stories of great events to shape public opinion,

   3. **Potential Practical Significance:** in order to help readers protect themselves from unscrupulous politicians.

Most readers are likely to think that connection is a bit of a stretch.

To formulate a useful applied research problem, you have to show that the answer in step 2 plausibly leads to step 3. Ask yourself this question:

(a) If my readers want to achieve the goal of _____ [state your objective from step 3],

(b) would they think that a good way to do that would be to find out _____? [state your question from step 2]

Try that test on the applied astronomy problem:

> (a) If my readers want to use data from earthbound telescopes
> to measure more accurately the density of electromagnetic radia-
> tion,
> (b) would they think that a good way to do so would be to
> find out how much the atmosphere distorts measurements
> of it?

Since astronomers have piles of data from earthbound telescopes
that could be adjusted for atmospheric distortion, the answer
would seem to be *Yes*.

Now try the test on the Alamo problem:

> (a) If my readers want to achieve the goal of helping people pro-
> tect themselves from unscrupulous politicians,
> (b) would they think a good way to do that would be to find out
> how nineteenth-century politicians used stories of great events
> to shape public opinion?

Again, that feels like a stretch.

If you really think that the answer to your research problem
can apply to a practical one, formulate your problem as the
pure research problem it is, then *add* your application as a fourth
step:

1. **Topic:** I am studying the differences among various nineteenth-
century versions of the story of the Alamo
    2. **Question:** because I want to find out how politicians used stories
    of great events to shape public opinion,
        3. **Conceptual Significance:** in order to help readers under-
        stand how politicians use elements of popular culture to ad-
        vance their political goals,
            4. **Potential Practical Application:** so that readers can better
            protect themselves from unscrupulous politicians.

When you state your problem in your introduction, it's usually
best to formulate it as a purely conceptual research problem
whose significance is based on conceptual consequences. Unless

your assignment includes the question of practical applications, save them for your conclusion. (For more on introductions and conclusions, see chapter 14.)

Most research projects in the humanities and many in the natural and social sciences have no direct application to daily life. In fact, as the word *pure* suggests, many researchers value pure research more highly than they do applied. They believe that the pursuit of knowledge "for its own sake" reflects humanity's highest calling—to know more and understand better, not for the sake of money or power, but for the good that understanding itself brings. As you may have guessed, the three of us support both the pure and the practical—so long as the research is done well and is not corrupted by dishonest or malign motives.

A threat to both pure and practical research today, especially in the biological sciences, is that profits from patents not only determine the choice of research problems, but also color their solutions: *Tell us what to look for, and we'll provide it!* That raises the kind of ethical question that we touch on later (pp. 285–88).

A TYPICAL BEGINNER'S MISTAKE

For some beginners, especially in classes that study significant practical problems, research problems never feel practical enough, not even when they have obvious applications. So they try to force their project into the practical domain. That's usually a mistake. No one can solve the world's great problems in a five- or even a fifty-page paper. But a good researcher might help us understand those problems better, which gets us closer to a solution. So if you care deeply about a practical problem, such as the increasing frequency of highly destructive forest fires in the West, carve out of it a research question that you can answer and that might ultimately contribute to a practical solution:

> How important are fires to the ecological health of a forest? How do local fire codes affect the susceptibility of buildings to fire damage?

Choose one of the smaller questions, knowing that small answers to small questions sometimes lead to great solutions.

4.3 FINDING A GOOD RESEARCH PROBLEM

What distinguishes great researchers from the rest of us is the brilliance, knack, or just dumb luck of stumbling on a problem whose solution makes the rest of us see the world in a new way. We can all learn to recognize a good problem when we bump into it, or it bumps into us (or when it's already a live issue). But researchers often begin a project without being entirely clear as to what their problem is. Sometimes they hope only to define it more clearly. Indeed, those who find a new problem or manage to clarify an old one often win more fame and (sometimes) fortune than those who solve a problem already defined. Some researchers have even gotten credit for *disproving* a plausible hypothesis that they had hoped to prove. So don't be discouraged if you can't formulate your problem fully at the outset of your research. Few of us can. But thinking about it early will save you hours of work along the way—and perhaps avoid panic toward the end.

Here are some ways you can aim at a problem from the start.

4.3.1 Ask for Help

Do what experienced researchers do: talk to teachers, classmates, relatives, friends, neighbors—anyone who might be interested in your topic and question. Why would anyone need an answer to your question? What would they do with it? What questions might your answer raise?

If you are free to select your own problem, look for a small one that is part of a bigger one. Though you are unlikely to solve the big one, your piece of it will inherit some of its significance. (You will also educate yourself about the problems of your field, no small dividend.) Ask your teacher what she is working on and whether you can work on part of it. But a warning: If your teacher helps you define your problem and gives you leads on sources, do not let those suggestions define the limits of your research. Nothing discourages a teacher more than a student who does *exactly* what is suggested, *and nothing more.* In that situation, the teacher probably wants you to do some research that will help her find out something she didn't know or understand, such as better sources and new data.

4.3.2 Look for Problems as You Read

You can find a research problem if you read critically. As you read a source, where do *you* detect contradictions, inconsistencies, incomplete explanations? If you are not satisfied with an explanation, if something seems odd, confused, or incomplete, tentatively assume that other readers would or should feel the same way. Many research projects begin in an imaginary conversation that a researcher has with another's report: *Wait a minute, he's ignoring . . .*

But before you set out to correct a gap, error, or misunderstanding, be sure it is real, not just your own misreading. Reread your source carefully and generously. Countless research papers have aimed to refute a point that no writer ever made.

Once you think you have found a real puzzle or error, do more than just point it out. If a source says X and you think Y, you have a research problem only if you can show that those who go on believing X will misunderstand something even more important. (For the most common kinds of contradictions, see our Quick Tip, pp. 72–74.)

Finally, read the last few pages of your sources closely. That's where many researchers suggest more questions that need answers. The author of the following paragraph had just finished explaining how the daily life of the nineteenth-century Russian peasant influenced his performance in battle:

> And just as the soldier's peacetime experience influenced his battlefield performance, so must the experience of the officer corps have influenced theirs. Indeed, a few commentators after the Russo-Japanese War blamed the Russian defeat on habits acquired by officers in the course of their economic chores. *In any event, to appreciate the service habits of Tsarist officers in peace and war, we need a structural—if you will, an anthropological—analysis of the officer corps like that offered here for enlisted personnel.* [our emphasis]

That last sentence gives us both the problem that this writer set out to solve and a new one waiting for someone to tackle.

4.3.3 Look for the Problem that Your Claim Solves

Critical reading can also help you discover a good research problem in your own early drafts. Writers almost always do their best thinking in the last few pages of a draft. It is often only then that they begin to formulate a final claim that they did not dream of when they started out. If in an early draft you arrive at an unanticipated claim, ask yourself what questions it might answer. Paradoxical as it might seem, you may well find a solution to a problem that you have not yet posed. Your task is to figure out what that problem is. Chances are, you can work backward to formulate a better, more interesting problem than the one that got you started.

4.4 SUMMARY: THE PROBLEM OF THE PROBLEM

Your teachers will assume that you are not an expert researcher, but they want you to start developing and practicing the mental habits of one. They want you to do more than just accumulate and report facts about a topic that happens to interest you. They want you to formulate a question that you think is worth answering and pose a problem that you think is worth solving, regardless of who else cares.

Eventually, though, as you move to advanced work, you have to share your new knowledge and understanding with others. At that point, you must understand what *your readers* think are interesting questions and problems. As we've emphasized, they base that judgment on the costs *they* pay as a result of not knowing or understanding something. And the step we all dream of is not only finding the kind of problem readers want to see solved, but persuading them to think seriously about a problem none of them has ever thought of. No one takes all three steps the first time out. Just about all of us get to the first one: *What am I interested in discovering?* Most of us get to the second: *What might my readers be interested in?* Few of us get to the third: *How can I get them to realize they are asking the wrong questions?* But those of us who don't get there do not necessarily fail, because we can measure our success by how well our readers think we answer questions they already care about. The worst response you can get from a reader is not *I don't agree,* but *I don't care.*

By now, all this airy talk about academic research may seem disconnected from a world in which so many people labor so hard at getting ahead or keeping others down. But when research problems in the world are pursued honestly, they are structured *exactly* as they are in the academic world. And in business and government, in law and medicine, in politics and international diplomacy, no skill is valued more highly than the ability to recognize a problem that others should take seriously, then to articulate that problem in a way that convinces them to care. If you can do that in a class in Chinese history, you can do it in a business or government office down the street or in Hong Kong.

You discover the most common kind of research problem when you disagree with a source. We can't tell you what to disagree with in them, but we can list some standard contradictions. This list will be most useful if you are familiar with research in a field, but if you're new, they can show you the kind of contradictions that experienced researchers look for. In chapter 14 we explain how to use these contradictions to write an introduction that motivates your readers to read on. (This list is not exhaustive, and some kinds overlap. You can also try them out on your topic.)

CONTRADICTIONS OF KIND
You claim that something thought to be one kind of thing is not (or vice versa).

> Certain religious groups are widely considered to be "cults" because of their strange beliefs, but those beliefs are no different in kind from standard religions.

In the following frames, substitute for X and Y terms of your own. In each case, you can also assert the opposite (that is, though X seems *not* to be a Y, it really is).

1. Though X seems to be a Y, it is not.

2. Though X seems to be a necessary characteristic or quality of Y, it is not.

3. Though X seems to be good/significant/useful/beautiful/ moral/interesting/ . . . , it is not.

PART-WHOLE CONTRADICTIONS
You claim that others mistake the relationship among the parts of something.

In recent years some have argued that athletics has no place in education, but in fact athletics is an intrinsic part of a well-rounded educated person.

1. Though X seems not to be a part of Y, it is.

2. Though part X seems to relate to part Y in Z way, it does not.

3. Though it is claimed that all X's have Y as a part, they do not.

DEVELOPMENTAL/HISTORICAL CONTRADICTIONS

You can claim that others have mistaken the origin, development, or history of your object of study.

Although some have recently argued that the world population is rising, it is not.

1. Though X seems to be stable/rising/falling . . . , it is not.

2. Though X may seem to have originated in Y, it did not.

3. Though the sequence of development of X seems to be 1, 2, and 3, it is not.

4. Though X seems to be part of a larger historical development, it is not.

EXTERNAL CAUSE-EFFECT CONTRADICTIONS

You can claim that assumed causal relationships do not exist (or vice versa).

A new way to stop juveniles from becoming criminals is the "boot camp" concept. But evidence suggests that it does little good.

1. Though X seems to cause Y, it does not.

2. Though X seems to cause Y, both X and Y are caused by Z.

3. Though X and Y seem to be causally correlated, they are not.

4. Though X seems to be sufficient to cause Y, it is not.

5. Though X seems to cause only Y, it also causes A, B, and C.

CONTRADICTIONS OF PERSPECTIVE

These contradictions run deeper. Most contradictions do not change the terms of the discussion. In perspectival contradictions, the author suggests that everyone must look at things in a new way.

> It has been assumed that advertising is best understood as a purely economic function, but in fact it has served as a laboratory for new art forms and styles.

1. X has been discussed in Y context, but a new context of understanding reveals new truth about X . . . (The new context can be social, political, philosophical, historical, economic, academic, ethical, gender specific, etc.)

2. X has been used to explain Y, but a new theory makes us see it differently.

3. X has been analyzed using theory/value system Y, leading to a rejection of X as inapplicable to Y. But now we see that Y is relevant to X in a new way.

CHAPTER FIVE

From Problems to Sources

If you are a beginning researcher and expect to find most of your data either in your library or on the Internet, use this chapter to develop a plan for your research. If you are more experienced, you may want to skip to the next chapter. If you are very experienced, skip to part III.

If you have not yet formulated a clear research question, you will have to spend some time reading around just looking for a topic that you can narrow down and question, as we described in chapter 4. But if you have a question and at least one candidate for a plausible answer (the philosopher C. S. Peirce had an apt name for it: a *hypothesis on probation*), you can start looking for data to test it. That doesn't mean lining up all the sources you can find and plowing through them to see what turns up. You want to look for *reliable* sources whose data let you test your hypothesis because they support it or, more importantly, challenge you to alter or abandon it.

If, however, you plunge into a search for sources without a plan, you risk losing yourself in a morass of books and articles. Sources can lead anywhere and everywhere, so it is easy to wander aimlessly from one to the next. To be sure, aimless browsing can be fun: everyone who loves learning loves to wander from book to book, idea to idea. Browsing can also be surprisingly productive: many important discoveries have been made through a chance encounter with a new idea that no one could have deliberately looked for. So we don't condemn all aimless reading; the three of us do it a lot.

But if you are working to a deadline, you don't have time to rely on chance: you have to search deliberately. In this chapter we'll talk about the resources you can look for and how to narrow them to a manageable list. In the next, we will discuss how to work with your resources once you find them. But as we've said, don't expect a linear plan that gets you from start to finish with no detours. You'll loop back as often as you move ahead. Just keep in mind that you are screening sources for data, arguments, and views that either confirm your hypothesis or give you reason to reject it.

> ### THREE KINDS OF SOURCES
>
> PRIMARY SOURCES: These are the materials that you directly write about, the "raw data." In fields like history and literature that study writers and documents, primary sources are texts from the period or by the author you are studying. In such fields, you can rarely write a research paper without using primary sources.
>
> SECONDARY SOURCES: These are research reports, whether books or articles, based on primary data or sources. You can quote or cite them to support your own research. If a researcher quoted your research report to support his argument, your report would be his secondary source. If, on the other hand, he were writing your biography, your paper would be a primary source.
>
> TERTIARY SOURCES: These are books and articles based on secondary sources. They synthesize and explain research in a field, usually for a popular audience. Generally, they just restate what others have said. Tertiary sources can help in the early stages of research, when you are trying to get a sense of a whole field, but they are weak support for new claims because they usually oversimplify, are seldom up-to-date, and are consequently mistrusted by most experts.

5.1 SCREENING SOURCES FOR RELIABILITY

Your question and hypothesis give you your most important basis for screening sources: they help you focus only on those that test your hypothesis, either supporting it or challenging it. If a source is on topic but irrelevant to your hypothesis, it may be interesting but it won't be immediately useful.

As you screen for relevant sources, you should also apply a second test: *Is this source reliable?* Just as one relevant source is more valuable then a dozen irrelevant ones, so one reliable source is more valuable than a dozen that are unreliable. As you look for sources, focus first on those you can trust.

There is no formula for testing the reliability of a source. But unless you are a very advanced student, you can usually rely on a few indications of reliability. Advanced researchers are expected to check the reliability of sources for themselves, but beginners will satisfy most readers if their sources have at least one of these characteristics:

- The source is published by a reputable press.

Most university presses are reliable, especially if it's a university whose name you recognize. In some fields, commercial presses have a reputation as strong as university presses, presses such as Norton in literature, Ablex in sciences, or Westlaw in law.

- The publisher uses peer reviews for everything it publishes.

You have no better guarantee of the reliability of a publication than its having been reviewed and approved for publication by independent experts in the field. Most books from reputable presses are peer-reviewed, though many essay collections are reviewed only by the named editor(s). The best scholarly journals require peer review, but some good ones do not.

- The author is a reputable scholar.

Books and journals usually tell you something about the credentials of the author, and you can easily find out more on the Internet.

- The source is current.

You must use up-to-date sources, but what counts as current depends on the nature of the source and field. In computer science, articles can be out-of-date in months. In philosophy, primary sources are current for centuries, secondary ones for decades. In general, a source that sets out a major position or theory that most other researchers accept will stay current longer than those that respond to or develop it. Assume that most textbooks are *not* current (excepting, of course, this one).

For secondary works, you can gauge the standards for currency by looking at journal articles in the works cited: What is the oldest date? Where do the dates tend to cluster? For primary works (novels, plays, letters, etc.), try to find out what is considered the standard edition; sometimes older editions are trusted more than more recent ones.

These indicators do not guarantee reliability. Reviewers sometimes recommend that a reputable press publish something weakly argued or with shaky data because other aspects of its research are too important to miss—we three have each done so. So don't assume that you can read uncritically everything written by a reputable researcher and published by a reputable press. In chapter 6 we'll talk about critical reading and in chapter 9 about evaluating the data you find in a source. But for a start, these indicators recommend a source as worth considering.

You can get a quick take on the most reliable sources on your topic by consulting the bibliography at the end of this book or by looking at one of the guides to research in your field (also listed there). Once you have located one reliable academic book or article, you have a trailhead for finding more: its footnotes and works cited point to sources you can track down, and their citations will point still farther down the trail.

WHOM CAN YOU TRUST?

According to a review committee appointed by the *Journal of the American Medical Association*, one of the more respected medical journals, "many statistical and methodological errors were common in published papers," even though those papers had been reviewed by experts in the field ("When Peer Review Produces Unsound Science," *New York Times*, June 11, 2002, p. D6). Some of you might just want to throw up your hands and give up on the idea of reliability: if the hyper-careful review procedures of *JAMA* don't guarantee reliable data, what's a mere student to do? You do what we all do—the best you can. Read critically, and when you report data, do so as accurately as you can. We'll return to this question in chapter 8.

5.2 LOCATING PRINTED AND RECORDED SOURCES

Unless you are collecting data from experiments or observation, you will probably find your data in books or articles, occasionally in photos and films or video and audio recordings. Your first stop should be your school library or a public library. You may even find one that specializes in your topic, such as the seventeenth-century collection at the W. A. Clark Library at UCLA; in a cause, such as the National Rifle Association Library in Fairfax, Virginia; or in a person, such as the Martin Luther King library in Atlanta and the many presidential libraries.

If the libraries near you are small and lack books and journals on your topic, start your research early so that you have time to borrow those you need through interlibrary loan. But no matter how small, your library probably offers more help than you suspect, including reference works, both general and specialized, research guides, and a variety of catalogs, bibliographies, and databases.

A caution: Some Internet-savvy students think that the best way to start their research is to enter their topic into a search engine and see what turns up. That can be a good way to find material out of which you can formulate a research question, but it is a very bad way to find reliable sources. Begin your search with your library: its catalog, bibliographies, and databases, which you may be able to access on the Internet.

5.2.1 Librarians

If you know your library, look for sources. But if this is your first shot at serious research, you might first talk to a librarian. Librarians are usually eager to help when you don't know where to start. Many libraries have special reference librarians, and large ones even have specialists in particular topics. They can show you how to use the online catalog, essential knowledge these days for any researcher. If you feel too shy or proud to ask, find out whether your library has e-mail service for reference questions. Otherwise, just go talk to a librarian.

The most important work you can do this early in the process is to *plan*. If you aren't ready to be helped, no librarian can help you. You will save your time and not waste hers if you prepare questions. Start by describing your project: try using the three-step rubric in chapter 3 to formulate an "elevator story" summarizing what you plan to do:

> I am working on the topic of _____, so that I can find out _____, because I want my readers to understand better _____.

Early on, your questions may be general: *Which periodical guides list articles about educational policy in the 1950s?* But as you narrow your topic, frame questions so that your librarian can understand exactly what you need: *How do I find court decisions on the "separate but equal" doctrine in educational policy in the early 1950s?*

5.2.2 General Reference Works

If you know a lot about your topic, focus on the specific sources you'll need. If not, start with general reference works such as the *Encyclopaedia Britannica* or with specialized ones such as the *Encyclopedia of Philosophy*. They give a reliable overview of your topic, and at the end of the article, they usually provide a list of sources that constitute the basic texts in a field. If you find nothing under one heading, look under another one. For example, the 1993 *Books in Print* listed nothing under *gender,* the term that is now standard for researchers in women's studies, but it had many entries under *sex.*

One new graduate student at the University of Chicago needed three trips to find where its research library keeps most of its books. She spent two trips wandering through seven floors of reading rooms, finding only reference works. Only on the third day did she get up enough nerve to ask a librarian where all the books were. She was directed to a door that led into the main stacks. Moral of the story: Ask!

5.2.3 Specialized Reference Works

Most fields provide extensive bibliographical resources, both in print and online. Large libraries offer online access to bibliographical databases covering most fields, many of them including abstracts. In some newer or highly specialized fields, you may find bibliographical lists on websites maintained by individual scholars, by departments, or by scholarly associations. These may be less reliable than the large databases, but they can get you started.

You should also find print bibliographies covering your whole field or specific aspects of it. If you are lucky, you'll find an annotated bibliography that briefly describes current books and articles; it is one of the best ways to get a quick overview of what other researchers think is important. Most fields publish a journal that reviews new research annually, which is even more useful. If you need the most current sources, the *Chronicle of Higher Education* regularly lists new books, and many journals list "books received" (new books that publishers send hoping the journal will review them).

5.2.4 Research Guides

Every major field has at least one guide to the resources that experienced researchers commonly use: lists of bibliographies, locations of important primary materials, research methods, and so on. Depending on how much time you have, you may want to look carefully at such guides, particularly if your library holds materials that the guides cite. The first step in learning the ropes of research is to find out where the ropes are stored. (We've listed some of the more popular ones in our "Appendix on Finding Sources.")

5.2.5 The Library Catalog

You can start with a general keyword search in your library catalog, but you'll work more efficiently if you first check bibliographies for specific titles. If your library does not have a title you need, request it through interlibrary loan (if you start early enough).

Once you locate a few sources you consider reliable, you can expand your search in two ways: keyword searches and browsing.

For a keyword search, start by entering specific terms in the titles you have already found—for example, *Alamo, Texas independence, James Bowie.* To expand your search, look for subject headings in the bibliographical data for each title (they may be on a "details" page online, or if you have the actual book, on the back of its title page). Those headings are the Library of Congress categories for all books. A search for them will generate titles related to your question, but also many that are not.

A quick way to expand on a small catalog is to consult the online catalog of the Library of Congress (www.loc.gov). It is easy to search, and you'll find almost any book, film, or recording you could want. It also has links to many university library catalogs. Your library may have only a fraction of what you find there, but it can borrow most of what you need. For books too new to be in a library catalog, consult an online bookseller. Those books you'll probably have to buy.

Be aware that if your library is large or you use the Library of Congress catalog, a keyword search can generate a vast number of titles. The University of Chicago library has almost three hundred books on Napoléon, and more than three thousand books with the word *environment* in their title. If your search turns up too many titles, narrow the list using the techniques we talked about in chapter 3.

On the other hand, if you exhaust the terms you can think of and still find nothing, your topic may be too narrow or too far off the beaten track to yield quick results. Or you could be onto an important question that nobody has thought about before, or at least not for a long time. Centuries ago, for example, "friendship" was an important topic for philosophers, but then was dropped and long ignored by major encyclopedias. Recently, though, it has been revived as a serious topic. In either case, chances are you'll make something of your topic only through your own hard thinking. In the long run, your topic might make you famous, but it is not one for a paper due in a few weeks.

5.2.6 In the Stacks

The second way to expand your list is to do some of the casual browsing we've recommended. If you can get into the stacks, skim the titles of books shelved on either side of those on your topic (look first at books with the newest bindings). Many will be irrelevant, but you are likely to find some that shed surprising new light on your question. (All three of us have found invaluable sources in this way.) You may also be able to browse online. Many library catalogs allow users to browse not only by call number or shelf location but also by subject and author. If yours does not, try browsing the Library of Congress catalog.

Finally, our advice assumes that your library has an online catalog. If yours does not or if its catalog is only partly online, you can do most of the tasks we've outlined with a card catalog, though more slowly. But we recommend that you start not with the cards, but online with the Library of Congress. A few quick searches there will give you an overview of what you might find in your library or borrow through interlibrary loan.

5.2.7 Online Databases

If most of your sources are not books but journal articles, skip the catalog and go right to your library's online databases. Although their search capabilities vary, most let you search for titles and keywords in all the ways we've described. (Browsing capabilities, however, are rare.) In addition to bibliographical data, many databases also include abstracts, which can speed the process of deciding which articles are worth reading carefully. Some databases even provide the full text of articles, though often for a fee. For information too current for the journals, check periodical indexes, or search the online archives of a major newspaper.

5.3 FINDING SOURCES ON THE INTERNET

The Internet changes so fast that generalizing is risky. Here is a principle that is true today, but may not be tomorrow: *Unless you have good reason not to, prefer a printed source to one on the*

Internet. (Not wanting to walk to the library is not a good reason.) Although you should never trust any source blindly, most of your readers will be more willing to trust print sources from reliable presses or journals than almost any source on the Internet.

Print sources are more highly respected because most of the data you can find on the Internet are not reliable enough for serious research. What complicates this generalization is that every day the Internet gains information as reliable as the best print data. You can find rigorously edited online journals, moderated discussion lists whose reviews and other edited contributions offer reliable scholarship, editions of primary texts superior to the best printed ones, and much more that is reliable. But such Internet-based sources stand beside incompetently edited journals, discussion lists full of nonsense, some of the least reliable editions of primary texts, and other data that are biased, distorted, invented, or simply the ravings of a demented mind.

The strength of the Internet is also its problem: it has no gatekeepers. It is like a publishing house without editors or a library without librarians. Consequently, you have access to more than the publishers or librarians provide, but you bear the risk of not knowing what parts of it are worth reading, can be trusted, have been checked for errors, and so on. So avoid using an Internet source unless you know that it is reliable and can persuade your readers to think so too. And never rely on the Internet to have a balanced or complete selection of sources. For the most part, people post what they are passionate about, so not only are individual postings liable to be biased, the selection is almost certain to be.

On the other hand, there are some situations in which you can use the information you find on Internet sources reliably:

- It is provided by a reliable journal or online publisher.

- It is in precisely the same form you would find in a library. Many government, civic, and business reports are released simultaneously on the Internet and in print.

- It supplements print sources. Some journals use the Internet to archive data not included in articles, to disseminate illustrations too expensive to print, or to host discussions between authors and readers.

- It is too recent to be found in libraries.

- It is available only on the Internet. Many government and academic databases are now available only online.

- It is your primary source. What is posted on the Internet is primary data about what people are thinking, the views of specific groups, and so on.

But remember: Before you treat a posting as reliable, evaluate the credentials of the poster and those who own, maintain, and sponsor the site.

To locate Internet sources, use the same techniques described for library catalogs, this time on a search engine. You won't find lists of subject headings, but you can use the same ones you used there. Be prepared, however, to pick through a lot of dross. (In our "A Note on Some of Our Sources," we cite some guides to research on the Internet. They offer more detailed advice about Internet-based research than we can offer here.)

Caution: You can find many printed texts posted on the Internet in violation of the author's copyright. Careful readers mistrust unauthorized copies because they are so often inaccurately reproduced. Ethical readers dislike seeing them cited because they violate the law. So unless a text is clearly posted with the author's permission (as in a database), use the printed rather than the Internet version of the text.

5.4 GATHERING DATA DIRECTLY FROM PEOPLE

Most projects can be done from books, journals, and the Internet alone, but you may also need data available only from talking with people. And again, the most important work you can do before you consult them is to *plan*. You will save time if you prepare specific questions. Help your source understand what you are up to by turning the three-step rubric in chapter 3 into an elevator story, ending with how you hope the person can help you:

I am working on the topic of _____, so that I can find out _____, because I want my readers to understand better _____. What I am hoping to learn from you is . . .

5.4.1 Experts as Sources of Bibliography

At every stage of research, you can usually find someone to guide you. At first, your teachers will help you focus your question and find sources. Here, too, the quality of the help you get depends on the quality of the questions you ask. The more you prepare before you talk to your teachers, the better you can explain what you are doing and the more they can help. Your teachers will not have all the answers, so you may have to look for help from others. (You might even hope that your teachers don't have all the answers, because then you will have something to teach them, and they will read your report with interest.)

You can never predict how much help you will need. At one extreme, we know a graduate student who met with his adviser every day for breakfast, reporting what he had found the day before and receiving guidance for the day ahead of him. (It's probably a good thing students rarely get that much help.) At the other extreme are those independent scholars who disappear into the library and never talk with anyone until they emerge with their project completed, sometimes years later. (We don't actually *know* any, but we hear they exist.) Most researchers choose a middle way, relying on casual conversations to guide their reading, which stimulates more questions and hunches to try out on others.

5.4.2 People as Primary Sources

In some areas, you have to collect primary data from people. Even if your research is not directly about individuals, you may still find people willing to provide information, if you can help them understand your interest in what they know. Don't ignore people in local industrial, governmental, or civic organizations. For instance, if you were researching school desegregation in your town, you might read court cases concerning the "separate but equal" doctrine that your reference librarian helped you locate, but you might also ask the local school district whether anyone there has memories she or he would share.

We cannot explain the complexities of interviewing, but remember that the more you sort out what you know from what you *want* to know, the more efficiently you will get what you need. In short, plan. You don't need to script an interview around a set list of questions—in fact, that's a bad idea because it tends to freeze the interviewee. But prepare so that you don't talk to your source aimlessly. You can always go back to a book you have misunderstood, but people are usually not sources that you can return to repeatedly just because you did not prepare well enough to get what you needed the first time.

THE ETHICS OF USING PEOPLE AS SOURCES OF DATA

In recent years our society has become increasingly aware that when researchers study people, they may inadvertently harm them—not just physically but emotionally, by embarrassing them or violating their privacy. So every college or university now has a Human Subjects Committee that reviews all research directly or indirectly involving people, when done by students or professional researchers. Its aim is to ensure that researchers follow the maxim that should govern research as much as it does medicine: *First, do no harm.* So consult with that committee if you use people as sources of data—by interviewing them, surveying them, perhaps even just observing them. You don't need clearance if you informally talk with a few dorm mates for a paper in a first-year writing class (as a courtesy, you should still tell them what you intend to do with the information they give you). You will likely need clearance if you are an advanced undergraduate and want to circulate a survey in your dorm that collects personal data of any kind. But if you are an advanced researcher, you must without fail get clearance before you do any kind of research that involves people. Jumping through these hoops may feel like bureaucratic make-work, but if you don't, you could harm those who help you in ways you don't anticipate and your institution could pay a price.

5.5 BIBLIOGRAPHICAL TRAILS

When you find a book that seems useful, skim its preface. It may list the author's friends and family, but also those who the author thinks have done good work. Next, skim the works cited and index. The works cited section lists books and articles on the same or related topics, and the index will show which were used most often (generally, the more pages devoted to an author or book, the more important it is). Articles usually begin with a review of previous research, and most supply references.

Now comes the second round. If your list is short, read it all. If it is long and you need to shorten it, start with sources mentioned most often by the works you read in the first round. Focus on works most relevant to your problem, but don't ignore a work that is not mentioned but is on your topic—you will get credit for originality if you turn up a good source that few others have found. By following this bibliographic trail, you can find your way through even the most difficult research territory, because one source always leads to others.

5.6 WHAT YOU FIND

Among these resources, you may find some titles right on your topic. You may even feel a flash of panic when you discover an article whose title could have been yours: "Transforming the Alamo Legend: History in the Service of Politics." At that moment you might think, *There goes my project, nothing new to say.* You could be right, but probably not. Study the source to see if it settles *your* question. If it does, you have to formulate a new one. But when you see how your topic has already been treated, you will probably find something to say about that treatment. In fact, once you see how someone else has addressed your topic, you can usually do it better. If the author has failed to get things quite right, you have found unwitting help in formulating your problem and the gist of the introductory paragraphs of your report (see pp. 72–74).

The most important thing you can do at this stage of your research is to keep your research question at the front of your mind. You must screen sources for reliability, but you must also screen them for relevance: Do they look as though they will help you answer your question? Or even clarify it? If you have time, skim sources that are just "about" your topic, because you will surely find some of them useful. The trick is to read with an open mind, as omnivorously as your time allows, but with a mind that also can weed out those sources that do not speak to your specific question and its possible answers.

CHAPTER SIX

Using Sources

To make your research as reliable as you expect your sources to be, you have to use them fairly and accurately. In this chapter we explain how to read and take notes so that readers can trust you when you cite, rely on, or critique a source.

How you use the sources you find depends on where you stand in your search for a problem and its solution. If you have only a topic, you may have to do a lot of unfocused reading to find a question to pursue. Be alert for matters that spark some special interest, for things that surprise you, especially for claims that you find odd, puzzling, dubious, even wrong. If you can find something that *you* find worth pursuing, you are more likely to sustain an interest in your project and communicate that interest in your report.

If you intend to use the sources you have found to answer a question you have, then you can use your sources to test and support your answer. At this point, you have to analyze the arguments of your sources critically yet fairly and to identify data that you might use. At the same time, you have to record not only your own thoughts, responses, and analyses, but details from the source itself, all in ways that are accurate and easy to recover later. Those are skills highly valued not just in the classroom, but in every workplace as well.

The problem is, human nature works against you, in two ways. First, taking good notes requires discipline. When you hunt down support for your claim, you focus on finding, not recording information. So taking notes feels like a distraction from the main goal. In that circumstance, too many of us take notes in a shorthand that seems good enough at the time, but is not much use later—just ask Doris Kearns Goodwin, a prominent historian and TV pundit whose reputation was damaged by mistakes she attributed to not taking careful notes.

More important, once we come up with a hypothesis to test, most of us embrace it too strongly. As a result, we don't read sources as objectively as we should. When you seek to support a particular answer, you quickly spot data and arguments that confirm it, but you'll be tempted to overlook or reinterpret data that contradict or even just qualify it. And when the data are ambiguous, you'll be tempted to resolve ambiguities in your favor.

You have to guard against those biases, both in your own work and in your sources. In this chapter we show you how to ensure that you use secondary sources as accurately, critically, and fairly as time—and human nature—allow.

6.1 THREE USES FOR SOURCES

Most researchers think of secondary sources only as providers of evidence. But you can also use them in another way: as models of argument, forms of analysis, and rhetorical moves used by those in your field. You can even use sources to find a good question to ask.

6.1.1 Read for a Problem

If you are having trouble formulating a problem or question, focus your reading to find one. Look for claims that puzzle you, that seem inaccurate or simplistic, or for data that others have ignored or not pursued. You can even borrow the general form of their questions. If a source you like asks a question about one historical figure, you might ask the same question about a related one. Skim conclusions to journal articles; researchers often point out at the end issues they have left unresolved or new lines of possible research.

This should be quick, serendipitous reading, sensitive to what sparks your interest and gets you thinking. Write as you read, but record only your general responses and ideas. If you come across data that you think might be important, just note where you found them without recording them in detail. You can't be sure what data you need until you know the question you'll address. But record bibliographical data exactly so that you can get back to the source easily.

6.1.2 Read for an Argument

In research, originality counts. Your teachers won't demand that you be entirely original, but they will reward you to the degree that you are. There is, however, one area where a research report is rarely original: in its logic. So one way to use a source is to borrow not its specific substance (that would be plagiarism), but the logic of its argument. (Academic argument is an issue we address in part III.)

Suppose you want to argue that the Alamo legend grew because it served the political interests of those who created it and satisfied the emotional needs of those who read or heard and repeated it. You will need reasons and evidence uniquely relevant to your claim, but readers will expect you to address the same *kinds* of points they look for in similar arguments about historical legends, real or fictional. They will expect you to say who created the legend and why; how the story was manipulated; whether the manipulation was deliberate; and so on. When you see how other researchers address similar problems, you can learn how to address yours in particular.

So if you have never made an argument like the one you think you may have to, find similar ones to use as models. When you take notes, record not the particular evidence but the larger claims; create an outline of the argument and note the *kind* of evidence used as support. It is likely to be the kind your readers will expect from you.

In your notes, turn each major point made by a source into a question to answer. If, for example, a source shows that creators of another legend benefited from responses to it, note that point and ask a corresponding question: *How did the Alamo legend benefit its creators?* Those questions can help you plan your outline.

You will probably not be able to touch on all the points in your sources, but they at least show you which ones readers are likely to look for.

Borrowing the *logic* of a source is not plagiarism. So long as you rely on a source only as a checklist of kinds of points to cover, you are not obliged to cite it in your text or works cited. You can, however, cite it (and gain some credibility) by observing that it makes an argument similar to yours:

> As Weiman (1998) has shown with regard to the Arthurian legends, those most responsible for the Alamo legend also gained the most from its depiction of Texas as an outpost civilization. . . .

In contrast to speedy reading that you do when looking for a question, reading for argument—or evidence—must be more careful. You must read slowly to get a sense of the whole argument in its complete context. A common cause of misunderstanding and misquoting is piecemeal reading—what is more aptly called "raiding." If you expect to use an argument or an idea, especially if you intend to quote it, read everything around it and anything else that you need to understand what you expect to use.

6.1.3 Read for Evidence

This is the most common reason for consulting sources: to find data useful as evidence to support a claim. When you find evidence, report it as completely and accurately as possible and cite the source fully, not only to give credit but to help readers find your source so that they can check the data for themselves. If you come across secondhand data (data that your source reports from another source), do everything you can to locate the original source. Not only can you then be sure your report is accurate (you may be surprised to see how often secondhand sources are not), but you may find other data equally useful. It is intellectually lazy and usually risky not to look up an important quotation in its original form and context, if that source is obtainable. (We'll return to reports of evidence in chapter 9.)

> Use statistical data only if you understand how to report them fairly and can also judge for yourself whether they were collected and analyzed appropriately. You will serve yourself well if you take courses in quantitative analysis, an area of knowledge of which most Americans are shamefully ignorant.

You don't have to agree with the conclusions in a source to use its data; in fact, its argument does not even have to be relevant to your question, so long as its data are. If you do find a source that makes a claim useful to you, you can cite it to support your own. But don't think that the *claim* is a fact you can use as data. All that claim shows is that another researcher agrees with you; to use it, you have to report not only that conclusion but its supporting data as well. (Of course, if a source makes exactly your claim, you must either find a new direction or frame your report as "further confirmation of Smith's claim.")

Don't try to collect all the data relevant to your question; that is usually impossible. But you do need sufficient and representative evidence. That can be hard to judge because different fields judge what is sufficient and what is representative differently. For example, to have sufficient evidence for a claim about a correlation between baldness and personality, a psychologist might need results from hundreds of subjects in many experiments. But before accepting a new cancer drug, the FDA might demand data from thousands of subjects through years of trials. The more at stake, the higher the threshold of sufficiency.

What counts as representative depends on the nature of the data. Anthropologists might interpret a whole culture in New Guinea on the basis of a deep acquaintance with a few individuals, but no sociologist would make a claim about American religious practices based on data from a single Baptist church in Oregon. If you cannot tell from your reading what your field judges to be sufficient and representative, consult your teacher or another expert. In particular, ask for examples of arguments that failed because their evidence was insufficient or unrepresentative. You learn what counts as right by accumulating representative examples of what goes wrong.

6.2 READING GENEROUSLY BUT CRITICALLY

When you read, be generous. Read first to understand fully. Go slowly; reread passages that puzzle or confuse you. If you cannot summarize a passage in your mind, assume you don't understand it well enough to use it in an argument. Don't start by assuming that you have to disagree with everything you find. In this first reading, resolve ambiguities in favor of the source. Prefer interpretations that help the source make sense, that make it more rather than less coherent. When a source presents an argument that may rival yours, you'll be especially tempted to read it in a way that emphasizes its weaknesses. Resist that temptation, at least at first.

But once you understand a source, you are free to disagree. Don't accept a claim just because an authority asserts it, especially when that assertion is not well supported. For decades people cited the "fact" that the Inuit peoples of the Arctic had lots of terms for types of snow. But when a researcher checked, she found that they have just three. (Or so she claims.) Be especially wary of dueling experts. If Expert A says one thing, B will assert the opposite, and C will claim to be an expert but is no expert at all. When some beginning researchers hear experts disagree, they become cynical and dismiss expert knowledge as mere opinion. Don't confuse uninformed opinion with informed and thoughtful debate.

Another aspect of critical reading is to check everything important for its accuracy (that's why we encourage you to chase down original data reported secondhand). Those who publish in respected journals rarely misrepresent their results deliberately. Yet if you ask almost any scholar whose work has been used by others, he will tell you that, as often as not, it has been reported inaccurately, summarized carelessly, or criticized ignorantly.

Publications like the Book Review of the Sunday *New York Times* or the *New York Review of Books* regularly print letters from angry authors responding to reviewers who, they claim, have misread or made factual errors in reviews of their books.

If you are unfamiliar with or can't find authoritative secondary sources—scholarly journals and books—you may have to resort to tertiary sources: textbooks, articles in encyclopedias, mass-circulation publications like *Psychology Today,* search engines like Google.com. If those are the only sources available, so be it, but never assume they are authoritative. Be especially wary of books and articles on complex issues aimed at mass audiences. It's not that journalists who write for ordinary readers about brains or black holes are necessarily incompetent; sometimes even distinguished researchers like the late Stephen Jay Gould write for popular audiences. But when they do, they always simplify, sometimes oversimplify, and their work usually dates quickly. So if you start your research with a popular book, look at the dates of the journals cited in its bibliography, then go to those journals, if you can, for the most current research.

Whom Do You Trust?

One of Booth's students got a summer job doing "scientific research" for a drug company. He was assigned to go through stacks of doctors' answers to questionnaires and shred certain ones until nine out of ten of those left did indeed endorse the company's product. The bogus files would then be used to "prove" that the product worked. The student quit in disgust and was, no doubt, quickly replaced by someone less ethically careful.

6.3 PRESERVING WHAT YOU FIND

Once you find sources that look promising, you have to read them purposefully and carefully, of course. In particular, you must record your sources in your bibliography accurately, and then when you take notes on them, you must record what you find accurately and fully.

6.3.1 Record *Complete* Bibliographical Data

Before you start taking notes, record *all* bibliographical data. We promise that no habit will serve you better for the rest of your career. For printed texts, record

- author,

- title (including subtitle),

- editor(s) (if any),

- edition,

- volume,

- place published,

- publisher,

- date published,

- page numbers of articles or chapters.

For online sources, record as much of the above information as applies: if you access a printed text online, you still have to cite the bibliographical information from the original printing. Also record

- URL,

- date of access,

- Webmaster (if identified),

- database (if any).

If you photocopy from a book, copy its title page; then write on it the publication date from its reverse side. Finally, record the library call number of the book or journal. You won't include call numbers in your works cited, but we can tell you how frustrating it is to find in your notes the perfect quote or the essential bit of data, whose source you incompletely documented. The call number will save you time when you have to go back to the library to recheck a source.

A few years ago, Williams had to withhold publication of some research on Elizabethan social structure for quite a while because he had failed to document a source fully. He had come across data that no one else had thought to apply to the problem he was addressing, but he could not use the data because he had failed to record complete information on the source. He searched the library at the University of Chicago for hours, until one night he woke up in bed, recalling that the source was in a different library!

6.3.2 Take Full Notes

When you are hunting down data, it can feel tedious to record them all accurately, but you can lose what you gain from reading carefully if your notes do not reflect the quality of your thinking. Some still believe that the best notes are written longhand on cards like this:

Sharman, <u>Swearing</u>, p. 133. HISTORY/ECONOMICS (GENDER?)

Says swearing became economic issue in 18th c. Cites <u>Gentleman's Magazine</u>, July 1751 (no page reference): woman sentenced to ten days' hard labor because couldn't pay one-shilling fine for profanity.

". . . one rigid economist practically entertained the notion of adding to the national resources by preaching a crusade against the opulent class of swearers."

(Way to think about swearing today as economic issue? Comedians more popular if they use bad language? Movies more realistic? A gender issue here? Were 18th-c. men fined as often as women?)

GT3080/S6

- At the top left of the card is the author, title, and page number.
- At the top right are keywords that help the researcher sort and re-sort cards into different categories and orders.

- The body of the card summarizes the source, records a direct quotation, and includes a thought about further research.

- At the bottom is the library call number for the book.

This format encourages systematic note-taking, but to be honest, we three no longer use cards (though we did a long time ago). We use a computer or a lined pad, because a note card is usually too small for what we want to write.

But we still follow these general principles:

- Put notes about different topics on different pages; don't jumble together on one page all your notes on different topics from a single source.

- On each sheet of notes record at the top the author, title, pages, and keywords. If you take notes on a computer, make them easier to search by using consistent keywords and shorthand titles.

- Perhaps most important: Clearly and unambiguously distinguish four kinds of references: what you quote directly, what you paraphrase, what you summarize, and what you write as your own thoughts. On a computer, use different fonts or styles; on paper use headings or different-colored sheets or ink.

However you take notes, be certain to record all the information you need to recover your critical reading and to let your readers know exactly how to find that same information.

6.3.3 Get Attributions Right

Here is why we stress distinguishing the words of your sources from your own. In recent years some eminent scholars have had their reputations shredded because they printed, as their own, the words of others that they had copied into their notes, but that they had "inadvertently" (they later claimed) failed to note were from the source. And we cannot emphasize too much that when you take notes, not only distinguish your own thoughts from those of the source, but also clearly and consistently distinguish summary, paraphrase, and direct quotes. Indicate direct quotations in a way you cannot mistake—large quotation marks, headings, a box around them, whatever you cannot overlook. The best way to distinguish the language of your source from your own and to ensure that your quotations are correct is to photocopy quotations longer than a few lines. Always record page numbers, not only of quotations and data, but of anything you paraphrase or summarize.

6.3.4 Get the Context Right

To support their claims, researchers build complex arguments out of several elements (we discuss them in detail in part III). As you assemble material from the arguments of your sources that you intend to use in yours, be aware of how they use their material.

1. **When you quote or summarize, be careful about context.** You cannot completely avoid quoting out of context, because you cannot quote all of an original. But if you read carefully and reread everything crucial to your own conclusions, you will draft summaries and copy quotations within the context that matters most, *the context of your own grasp of the original.* When you use a claim or argument, look for the *line of reasoning* that the author was pursuing and note it:

> NOT: Bartolli (p. 123): The war was caused by Z.

> NOT: Bartolli (p. 123): The war was caused by X, Y, and Z.

> BUT: Bartolli: The war was caused by X, Y, and Z (p. 123). But the most important cause was Z (p. 123), for three reasons: reason 1 (pp. 124–26); reason 2 (p. 126); reason 3 (pp. 127–28).

Sometimes you will care only about the conclusion, but experienced researchers never just add up votes—*Four out of five sources said X, so I do, too.* Readers want to see how your conclusions result from *arguments,* whether from your sources or your own. So when you take notes, record not only conclusions but also the arguments that support them. That way, you'll work in the context of *argued and related points.*

> Some misreporting happens because a researcher lazily relies on hearsay. Colomb heard a prominent researcher confess after her talk that she had never read an author whose work she had just discussed. One of Booth's books was "refuted" by a critic who apparently had read only the title of a section, "Novels Must Be Realistic." Failing to read beyond it, he didn't know that Booth himself was attacking the title, along with other misconceptions about fiction. One reviewer misquoted Williams and then, thinking he was disagreeing with him, used the misquoted evidence to argue for the point Williams originally made!

2. When you record the claim of a source, note the rhetorical importance of that claim in the original. Is it a main point? A minor point of support? A qualification or concession? A framing suggestion that is not a part of the main argument? By noting these distinctions you avoid this kind of mistake:

ORIGINAL BY JONES: "We cannot conclude that one event causes another just because the second follows the first. Nor can statistical correlation prove causation. But no one who has studied the data doubts that smoking is a causal factor in lung cancer."

MISLEADING REPORT ABOUT JONES: Jones claims that "we cannot conclude that one event causes another just because the second follows the first. Nor can statistical correlation prove causation." No wonder responsible researchers distrust statistical evidence of health risks.

Jones did not make that point at all. He *conceded* a point that he stated was relatively trivial compared to what he said in the final sentence, which is the point he really wanted to make. Anyone who deliberately misreports in this way violates basic standards of truth in research. But a researcher can make such a mistake inadvertently if he notes only the words and not their role in an argument.

Distinguish statements that are central to an argument from qualifications or concessions the author acknowledges but downplays. Unless you are reading a source "against the grain" of the writer's intention—for example, you want to expose hidden tendencies—do not report minor aspects of a research report as though they were major ones or, worse, as if they were the whole of the report.

Be especially attentive to "framing" statements at the beginning and end of an argument. Careful scholars usually frame their discussions with contextualizing statements. Sometimes those are their most interesting claims, but while they may believe them, they do not always support them.

3. Be sure of the scope and confidence an author expresses in making a claim. These are not the same:

Chemicals in French fries cause cancer.

Chemicals in French fries seem to be a factor in causing cancer.

Chemicals in French fries correlate with a higher incidence of cancer.

4. Don't mistake the summary of another writer's views for those of the author summarizing them. Many writers do not clearly indicate when they are summarizing another's arguments, so it is easy to quote those authors as saying the opposite of what they in fact believe and are actually setting out to disprove.

5. When dealing with sources that agree on a major claim, determine whether they also agree on how they interpret and support it. For example, two social scientists might claim that a social problem is caused not by environmental forces but by personal factors, but one might support that claim with evidence from genetic inheritance while the other points to religious beliefs. How and why sources agree is as important as the fact that they do.

6. Identify the cause of disagreement. Do sources disagree because they cite different evidence, because they interpret the same evidence differently, or because they approach the problem differently?

It is risky to attach yourself to what any one researcher says about an issue. It is not "research" if you just uncritically summarize another's work. Even if your source is a scholar who is universally trusted, be careful. If you rely on at least two sources, you will almost always find that they do not agree entirely, and that's where your own research can begin. *Which has the better argument? Which better* respects *the evidence?* In fact, there is a research problem right there—whom should we believe?

Finally, remember that your report will be accurate only if you double-check your notes against your sources. After your first draft, check your quotations against your notes. If you use one source extensively, skim its relevant parts. By this time, you may be seized by the enthusiasm we mentioned earlier. You'll *believe* in your claim so strongly that you will see all your evidence in its favor. Despite our best intentions, that temptation afflicts us all. There is no cure, save for checking and rechecking. And rechecking again.

For both beginners and experts, mistakes are part of the game. All three of us have discovered them in our published work (and desperately hoped no one else would). Mistakes are most likely when you copy a long quotation. When Booth was in graduate school, his bibliography class was told to copy a poem *exactly* as written. Not one student in the class of twenty turned in a perfect copy. His professor said he had given that assignment to hundreds of students, and perfect copies had been done by just three. But even when you make an especially foolish mistake, don't think you are the only one who ever has. Booth still winces when he remembers the graduate paper he turned in on Shakespeare's *McBeth,* and Williams would like to forget the report he was supposed to give in class, but never did, because he could find nothing on his assigned topic, that great Norwegian playwright Henry Gibson (it's *Macbeth,* of course, and Henrik Ibsen). In fact, until our very last proofreading, the story about Booth on page xiv had him standing before heaven's "Golden Gate."

6.3.5 Use Comments and Keywords to Organize Your Thoughts

Your notes should be faithful to your sources, but they should also reflect your own growing understanding of how you will explain and support your answer to an important question. So, as you take notes, start writing comments that reflect your thinking about how your data might fit into your argument. Regularly review them to take stock of where your argument is and how far it has to go. You can make that process easier if you use keywords.

Keywords name central concepts in your problem and its solution. Use some general keywords, such as *Alamo, politics, myth, history,* but concentrate on those that are specific to your particular argument: *outpost civilization, Mexican response, borderlands culture.* Select them carefully, especially if your project requires extensive research and you take notes on a computer. When you organize around thoughtful, specific keywords, you can search your computerized notes to combine and recombine them in novel ways. (If you type keywords with an asterisk—**outpost civilizations*—you can target your search more easily.)

6.4 GETTING HELP

As your research progresses, you may experience a moment when everything you have learned seems to run together. When that happens, you are probably accumulating data faster than you can handle them. You know a lot but can't be sure what's useful or relevant. You can't expect to avoid all such moments, but you can minimize the anxiety they create by taking every opportunity to organize and summarize what you have gathered *in writing* and *as you go,* and to keep returning to the central questions: *What problem am I posing here? What question am I asking? How are my data relevant to either?* Keep coming back to that formula, *I am working on X to learn more about Y, so that my readers can better understand Z.*

At moments of utter confusion, turn to friends, classmates, teachers—anyone who will serve as a sympathetic but critical audience. Explain how what you have learned bears on your question and moves you toward a resolution of your problem. Give your friends progress reports, asking: *Does this make sense? Am I missing an important aspect or question? Given what I have said, what else would you like to know?* You will profit from their reactions, but even more from the mere act of explaining your ideas to nonspecialists.

You owe readers a careful reading of an important source, but early on you may have to do some speed-reading to weed out useless ones. Successful speed-reading, however, requires more than just running your eyes down a page. To identify the main elements of an argument, you must have an idea of both the structure of the argument (see part III) and the geography of the book or article that reports it (see part IV).

1. **Become familiar with the geography of the source.** Before you skim, get a sense of the whole structure.

A. If your source is a book,

- read the first few sentences of each paragraph in the preface;
- look in its table of contents for prologues, introductions, summary chapters, and so on;
- skim the index for topics with the most page references;
- skim the bibliography, noting sources cited most often;
- flip through chapters to see if and how they are divided into sections with headings and if they have summaries at the end.

B. If your source is an article,

- read the abstract, if it has one;
- flip through to see if there are section headings;
- skim the bibliography.

2. Locate the problem/question and the solution/main claim.

A. If your source is a book,

- read introductions, summaries, and the first and last chapters.

B. If your source is an article,

- read the introduction, with special attention to its last two paragraphs, and the conclusion.

At this point, you may be able to rule out irrelevant sources. If not, do the following:

3. Identify key subclaims.

A. If your source is a book,

- read the first and last few paragraphs of each chapter;

- then read each chapter as if it were an article (see below).

B. If your source is an article,

- locate its sections;

- read the first and last paragraph of each section.

4. Scan for key themes. Start by scanning for key concepts. If you add those concepts to your notes on the bibliographical data, you can use them to help see connections among sources worth a closer look. If these steps point to a source that seems relevant, go back and *read it carefully,* a process that will be easier because you already have a sense of its important elements.

As you will see when we turn to planning and drafting, practice in this kind of speedy reading can help guide your own strategies of writing and revision. If your readers cannot skim *your* reports and discover the outlines of *your* argument, the organization of your own report will not have served them well.

PART III

Making a Claim and Supporting It

Prologue

PULLING TOGETHER YOUR ARGUMENT

If you have accumulated a bushel of notes, photocopies, and sum-maries, all spilling off your desk or filling up your hard drive, it's time to think about imposing some shape on all that stuff, especially if you can see even the dim outline of an answer to your research question. The risk, however, is that you may be tempted like too many researchers to sort your data under the most obvious topics, arrange them into some arbitrary sequence, and start writing. Unfortunately, the obvious topics are usually the least useful, because they will likely reflect only what your sources suggest. Even if those suggested topics do go beyond the obvious, they are likely to fit only a linear sequence (A + B + C + . . .), a structure usually too weak to support a complex argument. And almost surely they will not be organized in a way that clearly supports the claim that answers your question.

To impose a *useful* order on all that information, you need a principle of organization that comes not from the categories of your data but from the logic of your answer and its support. You have to organize your report to support a claim that answers your research question and justifies both the time you spent answering it and the time you ask readers to spend reading about it. The support for that answer and claim takes the form of a research *argument*.

Though you should at first organize your materials around the elements of your argument, your final draft must reflect not only the structure of your argument but also the structure of your readers' understanding. We will discuss these two steps as though you could take them separately: first assemble the elements of your argument and then arrange them to meet you readers' knowledge and needs. But the process of creating an effective report is cyclical, so as you focus on assembling your argument in part III, keep in the back of your mind our advice about planning a draft in part IV. As you become a more experienced writer, assembling your argument and planning your first draft will become a single action.

RESEARCH ARGUMENTS

In chapter 4 we distinguished everyday, troublesome problems from the kind that motivate research projects. In the same way, we now have to distinguish between everyday arguments and the kind that organize research reports. People usually think of arguments as disputes: children argue over a toy; roommates over the stereo; drivers about who had the right-of-way. Such arguments can be polite or heated, but they all involve conflict, with winners and losers. To be sure, researchers sometimes wrangle over evidence and occasionally erupt into charges of carelessness, incompetence, and even fraud. But that is not the kind of argument that made them researchers in the first place.

In the next five chapters, we examine a kind of argument that is less like a prickly dispute with winners and losers and more like a thoughtful conversation with amiable colleagues, a conversation in which you cooperatively explore a contestable issue that you all think is important to resolve, a conversation that aims not at coercing each other into agreement, but at cooperatively finding and agreeing on the best answer to a hard question.

In that conversation, though, you do more than just politely exchange opinions. We are all entitled to our opinions, and no law requires us to explain or defend them. But in a research community, we are expected both to make claims new and important enough to interest our readers and to explain them, as if our readers were asking us, quite reasonably, *Why should I believe that?*

In a research report, your goal is not to stuff your claim down your readers' throats, but to start where they do, with what they know and don't know, what they accept and what they question. Then you answer those questions in a way that lets readers see how your claim solves their problem, and so furthers their best interests. To do that, you must anticipate their questioning each element of your argument, not to knock it down, but to help you both find and understand a truth you can *share*. Of course, when you write an argument, they are usually not there to question you, so you must learn to imagine their questions so that your arguments truly are a conversation with readers.

GETTING TO KNOW YOU

Nothing is harder than imagining questions from someone you don't know. Experienced researchers have the advantage of knowing many of their readers personally. They talk with them, trying out ideas before writing them up. And when they don't know their readers, they try to find out about them.

A group of physicists who wanted biologists to notice their research were unhappy when the first manuscript they sent to a biology journal was rejected. So they started attending biology conferences, reading biology journals, even hanging around the lounge in the biology department. After they got to know how biologists think, they did some rewriting and were able to publish papers that influenced the field.

Students seldom have the time or opportunity to hang around their readers, especially before they start to specialize in a field. But you can do some homework on questions your readers might ask:

- Read journals that publish research like yours. Notice the kinds of questions the articles acknowledge and respond to.
- Rehearse your argument with your teacher. After you have a plan but before you draft, talk over your ideas, asking whether any seem confusing or doubtful to her.
- Ask someone to read your drafts and indicate where they have questions or see alternatives. Find someone as much like your intended readers as possible

You've been told a thousand times to think about your readers. To do that, you have to get to know them.

CHAPTER SEVEN

Making Good Arguments

AN OVERVIEW

In this chapter we discuss the five elements of research arguments, showing how they respond to readers' predictable questions and how you can organize them into a genuinely coherent argument.

When you know enough to start planning your research report, you should have a tentative but clear understanding of your question and why it might matter to your readers, and a tentative but reasonably specific answer. You should have a list of reasons that support your claim and evidence to support those reasons, and some idea about the kinds of questions and objections your readers would be likely to raise, were they there in front of you. You won't be able to imagine all of their questions, nor will they expect you to. But you must anticipate at least the questions that generate the five elements of an argument and answer them before they're asked.

7.1 ARGUMENT AND CONVERSATION

In a research report, you make a **claim**, back it with **reasons** based on **evidence**, **acknowledge** and **respond** to other views, and sometimes explain your **principles** of reasoning. There's nothing arcane in any of this, because you use those elements in every conversation that inquires thoughtfully into an unsettled issue:

A: I hear you had a rocky time last semester. How do you think this term will go? [*A poses a problem that interests her, put in the form of a question.*]

B: Better, I hope. [*B makes a claim that answers the question.*]

A: Why is that? [*A asks for a reason to believe B's claim.*]

B: I'll finally be taking courses in my major. [*B offers a reason.*]

A: Why do you think that'll make a difference? [*A doesn't see how B's reason is relevant to his claim that he will do better.*]

B: When I take courses I'm interested in, I work harder. [*B offers a general principle that relates his reason to his claim.*]

A: What courses? [*A asks for evidence to back up B's reason.*]

B: History of architecture, introduction to design.

A: But what about that calculus course you have to take again? [*A offers a point that contradicts B's reason.*]

B: I know I had to drop it last time, but I found a really good tutor. [*B acknowledges A's objection and responds to it.*]

A: But won't you be taking five courses? [*A raises another reservation.*]

B: I know. It won't be easy. [*B concedes a point he cannot refute.*]

A: Will you pull up your GPA? [*A asks about the limits of B's claim.*]

B: I should. I'm shooting for at least a 3.0, as long as I don't have to get a part-time job. [*B limits the scope of his claim and adds a condition.*]

If you can imagine playing the roles of both A and B, you will find nothing strange about assembling a research report, because every written argument, research or not, is built out of the answers to those same five questions that you must ask on your readers' behalf:

1. What do you **claim**?

2. What **reasons** support that claim?

3. What **evidence** supports those reasons?

4. Do you **acknowledge** this alternative/complication/objection, and how do you **respond**?

5. What **principle (warrant)** justifies connecting your reasons to your claim?

7.2 BASING CLAIMS ON REASONS

At the core of every research report is your claim, the answer to your research question, along with two kinds of support for it. The first support is at least one **reason**, a sentence or two explaining why your readers should accept your claim. We can usually join a claim and a reason with *because*:

> The emancipation of Russian peasants was an empty gesture *claim* **because** it did not improve the material quality of their daily lives. *reason*

> TV violence can have harmful psychological effects on children *claim* **because** those exposed to lots of it tend to adopt the values of what they see. *reason*

At this point, we have to pause to clarify some terms. We must distinguish *claims* in general from *main claims,* and both from *reasons*:

- As we will use the term, a *claim* is any sentence that asserts something that may be true or false and so needs support: *The world's temperature is rising.*

- A *main claim* is the sentence (or more) that your whole report supports (some call this its *thesis*). If you wrote a report to prove that the world's temperature is rising, the sentence stating that would be its main claim.

- A *reason* is a sentence supporting a claim, main or not.

These terms can get confusing, because a reason is often supported by more reasons, which makes that first reason a claim in its own right. In fact, a sentence can be *both* a reason *and* a claim at the same time, if what it states (1) supports a claim and (2) is in turn supported by another reason: For example,

> TV violence can have harmful psychological effects on children *claim 1* because **those exposed to large amounts of it tend to adopt the values of what they see** *reason 1 supporting claim 1/claim 2 supported by reason 2* Their constant exposure to violent images makes

them unable to distinguish fantasy from reality.*reason 2 supporting reason 1/claim 2*

Reasons can be based on reasons, but ultimately a reason has to be grounded on *evidence*.

7.3 BASING REASONS ON EVIDENCE

In casual conversation, we usually support a claim with just a reason:

We should leave*claim* because it looks like rain.*reason*

We don't ask, *What evidence do you have that it looks like rain?* (unless someone thinks he's a meteorologist: *Those aren't rain clouds; they're just . . .*).

When you address serious issues in writing, though, you can't expect readers to accept all your reasons at face value. Careful readers behave more like that would-be weatherman, asking for the evidence, the data, the facts on which you base those reasons:

TV violence can have harmful psychological effects on children *claim 1* because those exposed to large amounts of it tend to adopt the values of what they see.*reason 1 supporting claim 1/claim 2 supported by reason 2* Their constant exposure to violent images makes them unable to distinguish fantasy from reality.*reason 2 supporting reason 1/claim 2* **Smith (1997) found that children ages 5–9 who watched more than three hours of violent television a day were 25 percent more likely to say that most of what they saw on television was "really happening."** *evidence supporting reason 2*

At least in principle, *evidence* is something you and your readers can see, touch, taste, smell, or hear (or is accepted by everyone as just plain *fact—the sun came up yesterday morning*). It makes no sense to ask, *Where could I go to see your reasons?* It does make sense to ask, *Where could I go to see your evidence?*

For example, we can't see children adopting values, but we could see a child answer the question *Do you think that what you see on TV is real?* That somewhat oversimplifies the idea of "evidence from out there," but it illustrates the principle. (We'll discuss this distinction between reasons and evidence in more detail in chapter 9.)

We now have the core of a research argument:

> **Claim** *because of* **Reason** *based on* **Evidence**

7.4 ACKNOWLEDGING AND RESPONDING TO ALTERNATIVES

A responsible researcher supports a claim with reasons based on evidence. But thoughtful readers don't accept a claim just because you back it up with *your* reasons and *your* evidence. Unless they think exactly as you do (unlikely, given the fact that you are making an argument), they will probably think of evidence you haven't, interpret your evidence differently, or, from the same evidence, draw a different conclusion. They may reject the truth of your reasons, or accept them as true but deny that they are relevant to your claim and so cannot support it. They may think of alternative claims you did not consider.

In other words, your readers are likely to question *any* part of your argument. So you have to anticipate as many of their questions as you can, and then acknowledge and respond to the most important ones. For example, as readers consider the claim that children exposed to violent TV adopt its values, some might wonder whether children are drawn to TV violence because they *already* are inclined to violence of all kinds. If you think readers might ask that question, you would be wise to acknowledge and respond to it:

> TV violence can have harmful psychological effects on children *claim 1* because those exposed to large amounts of it tend to adopt the values of what they see. *reason 1 supporting claim 1/claim 2 supported by reason 2* Their constant exposure to violent images makes them unable to distinguish fantasy from reality. *reason 2 supporting reason 1/claim 2* Smith (1997) found that children ages 5–9 who watched more than three hours of violent television a day were

25 percent more likely to say that most of what they saw on television was "really happening." *evidence supporting reason 2* **It is conceivable, of course, that children who tend to watch greater amounts of violent entertainment already have violent values,** *acknowledgment* **but Jones (1989) found that children with no predisposition to violence were just as attracted to violent entertainment as those with a history of violence.** *response*

The problem all researchers face is not just responding to readers' questions, alternatives, and objections, but imagining them. (In chapter 10 we'll review questions and objections you should expect.)

Since no research argument is complete without them, we add acknowledgment/responses to our diagram to show that they relate to all the other parts of an argument:

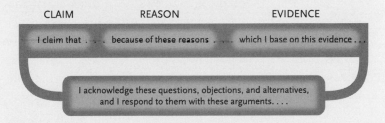

7.5 WARRANTING THE RELEVANCE OF REASONS

Even if readers agree that a reason is well supported by evidence, they may not see why it should lead them to accept your claim. They will ask why that reason, though factually true, is *relevant* to the claim. For example, suppose you offer this claim and its supporting reason (assume the evidence is there):

> Children who are exposed to large amounts of violent entertainment tend to become adults who think violence is a legitimate component of daily life *claim* because as children they tend to adopt the violent values in what they see. *reason*

Readers might question not the truth of that reason, but its *relevance* to the claim:

Why should children who adopt violent values necessarily become adults who tend to accept violence as a legitimate component of everyday life? I don't see how your claim follows from
your reason.

To answer, you must offer a general principle that shows why
you believe your *particular* reason is relevant to your *particular*
claim so that you are justified in connecting them:

Whenever children adopt particular values, as adults they tend to
accept as "normal" any behavior that reflects those values.

That statement—sometimes called a *warrant*—expresses a general principle of reasoning that covers more than violent TV. It
covers all values acquired as a child and all adult behaviors.

Think of a warrant as a principle claiming that a general set
of circumstances predictably allows us to draw a general consequence. You can then use that warrant to justify concluding that
a *specific* instance of that general consequence (your claim) follows from a *specific* instance of that general circumstance (your
reason). But for that warrant to apply, readers must first agree
that the specific circumstance (or reason) qualifies as a sound
instance of the general circumstance in the warrant and that the
specific consequence (or claim) qualifies as a sound instance of
the general consequence.

As you'll see, it is not easy to decide where to put warrants in
the sequence of an argument, or even whether you need them
at all. In fact, writers state warrants rarely, only when they think
readers might question the relevance of a reason to their claim.
For example, suppose you said:

Watch out going down the stairs, because the light is out.

You wouldn't need to add the warrant

When it's dark, you have to be careful not to misstep.*warrant* So
watch out going down the stairs,*claim* because the light is out.*reason*

That would seem condescending.

But if you think readers won't immediately see how a reason is relevant to your claim, then you have to justify the connection with a warrant, usually before you make it:

> Violence on television and in video games can have harmful psychological effects.*main claim* **Few of us question that when children are repeatedly exposed to particular values in graphic and attractive form, they use those values to structure their understanding of their world.***warrant* In the same way, children constantly exposed to violent entertainment tend to adopt the values of what they see. . . .

(As you can see, no aspect of argument is as abstract and difficult to grasp as warrants.)

We add warrants to our diagram to show that they connect a claim and its supporting reason:

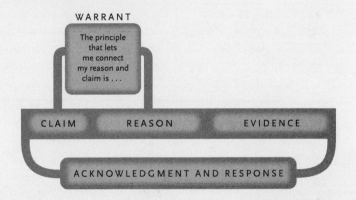

Those five elements constitute a "basic" argument. But many also include *explanations* of issues that readers might not understand. If, for example, you were making an argument about the relationship between inflation and various forms of money supply to readers not familiar with economic theory, you would have to explain the different ways that economists define "money."

7.6 BUILDING COMPLEX ARGUMENTS OUT OF SIMPLE ONES

The arguments in research reports are, of course, more complex than these simple ones. First, researchers almost always support a claim with more than one reason, each of which is supported by its own evidence and may be justified by its own warrant. Second, since readers can be expected to see many alternatives to any complex argument, careful researchers typically respond to a number of them.

But most important, each element of a substantial argument is itself likely to be treated as a claim, supported by its own argument. Each reason will typically be treated as a claim supported by other reasons, often reasons that are themselves claims. A warrant may be supported by its own argument, with reasons and evidence, perhaps even with its own warrant and acknowledgments and responses. Each response might itself be a mini-argument, sometimes a full one. Only the evidence "stands alone," but you may have to explain where you got it and why you think it's sound.

7.7 ARGUMENTS AND YOUR ETHOS

This process of "thickening" an argument with other arguments is one way that writers gain the confidence of readers. Readers will judge you by how well you manage the elements of an argument so that you anticipate their concerns. In so doing, they are in effect judging the quality of your mind, even of your implied character—an image of yourself that you project through your argument, traditionally called your *ethos*. When you seem to be the sort of person who supports your claims thoroughly and who thoughtfully considers other points of view, you give readers reason to trust what you say and not to question what you don't. By acknowledging their views and differences, you foster their desire to work with you in developing and testing new ideas.

In the long run, the ethos you project in individual arguments settles into your reputation, something every researcher must care deeply about, because your reputation will be an invisible sixth element in every argument you write. It answers the unspoken question *Can I trust this person?* If your readers don't know you, you have to earn that trust in each argument. But if they do know you, you want the answer to their question to be *Yes*.

In the next four chapters, we look at each element of an argument, to show you both how to assemble them into a complete argument and how to think about them critically. In part IV we take up the matter of arranging those elements into a coherent report.

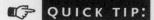 **QUICK TIP:** *Designing Arguments Not for Yourself but for Your Readers: Two Common Pitfalls*

Arguments fail for many reasons, but inexperienced researchers stumble most often when they rely too much on what feels familiar and comfortable and too little on what their readers need. Here are two common problems to avoid.

INAPPROPRIATE EVIDENCE

If you are working in a new field and unfamiliar with its characteristic modes of argument, you'll be tempted to fall back on forms of argument you already know. Every time you enter a new research community, though, you must find out what's new about the kinds of argument those in that community expect you to make. If you learned in a first-year writing class to search for evidence in your own experience or take a personal stand on issues of social concern, do not assume that you can do the same in fields that emphasize "objective data," such as experimental psychology. On the other hand, if as a psychology or biology major you learned to gather data, subject it to statistical analysis, and avoid attributing to it your own feelings, do not assume that you can do the same in art history.

This does not mean that what you learn in one class is useless in another. All fields share the elements of argument we describe here. But you do have to watch for what's distinctive in how a field handles those elements and be flexible enough to adapt—trusting, at the same time, the skills you already command. You can anticipate this problem as you read by noting the kinds of evidence used by the sources you consult. Here are just a few of the different kinds of evidence to watch for in different fields:

- personal beliefs and anecdotes from writers' own lives, as in a first-year writing course;

- direct quotations, as in most of the humanities;

- citations and borrowings from previous writers, as in the law;

- fine-grained descriptions of behavior, as in anthropology;

- statistical summaries of behavior, as in sociology;

- quantitative data gathered in laboratory experiments, as in natural sciences;

- photographs, sound recordings, videotapes, and films, as in art, music, history, and anthropology;

- detailed documentary data assembled into a coherent story, as in some kinds of history or anthropology;

- networks of principles, implications, inferences, and conclusions independent of factual data, as in philosophy.

Just as important, note the kinds of evidence that are never used in your field. Anecdotes enliven literary history but rarely count as good evidence in sociological explanations; fine-grained narratives are crucial in many anthropological reports but are irrelevant in an argument about subatomic physics.

COMFORTABLE SIMPLICITY

When you are new to a field, everything you read may seem confusing. Like everyone else in those circumstances, you will look for a familiar method or an unambiguous answer, any simplification that helps you manage the complexity. Once you find it, you are in danger of oversimplifying your argument. But no complex effect has a single unambiguous cause; no serious question has a single unqualified answer; no interesting problem has a single methodology to solve it. So when you are new to a field, seek out qualifications; formulate at least one alternative solution to your problem; ask whether someone else in the field approaches your problem differently.

As you learn the typical problems of a field, its methods,

schools of thought, and so on, you will begin to be comfortable with its standard forms of argument. It is at this point that newly experienced researchers succumb to another kind of overgeneralization: once you learn how to construct one kind of argument, you try to make that same argument over and over. Be aware that every field exhibits a second kind of complexity, the complexity of competing solutions, competing methodologies, competing goals and objectives—all marks of a lively field of inquiry. The more you learn, the more you recognize that while things are not as blindingly complex as you first thought, neither are they as simple as you then hoped they would be.

COGNITIVE OVERLOAD: SOME REASSURING WORDS

At this point, you may be feeling a bit overwhelmed. Take comfort in the fact that your anxieties have less to do with age or intelligence than with sheer lack of experience in a particular field. One of us was explaining to teachers of legal writing how being a novice makes new law students feel insecure. At the end of the talk, one woman reported that she had been a professor of anthropology whose published work had been praised for the clarity and force of her writing. Then she switched careers and went to law school. She said that during her first six months, she wrote so incoherently that she feared she was suffering from a degenerative brain disease. Of course, she was not: she was experiencing a kind of temporary aphasia that afflicts most of us when we try to write about matters we do not entirely understand for an audience we understand even less. She was relieved to find that the more she understood law, the better she wrote about it.

CHAPTER EIGHT

Claims

In this chapter we discuss the point of your argument, the claim that answers your research question and serves as the main point of your report.

As we have emphasized, you need a tentative answer to your research question well before you can know exactly what the final one will be. Even if you expect to replace your working answer, you need one from the start to help you know what to look for and to sift out from what you find just those data that are relevant. You also need that tentative claim to help you assemble the kind of argument you will need to support it. So from the first, try to articulate the best, most complete claim your current understanding allows.

You can test your claim with three questions:

- What kind of claim will you make?

- Can you state it specifically?

- Will your readers think it is significant?

When you can answer those three questions, you're ready to assemble other elements of your argument to see whether you in fact can make a good case for your claim.

8.1 WHAT KIND OF CLAIM?

The kind of problem you pose determines the kind of claim you make and the kind of argument you need to support it. As we saw in chapter 4, researchers in academic settings usually pose not a practical problem but a conceptual one, the kind whose solution asks readers not to *do* something but to *believe* something:

> The recession of 2001–2002 was caused partly by excessive investment in information systems that failed to improve productivity as much as had been promised.

Some conceptual claims might imply an action:

> Businesses that invest in information systems benefit only when they understand how to use them to improve productivity.

But if you want readers to act, it is wise to be explicit about what they should do: writers too often assume that readers can infer your intentions better than they actually do.

Some researchers think that by posing and answering a conceptual question, they can contribute to the solution of a practical problem: If we could simply *understand* what turns cancer cells on, we might figure out how to turn them off. But if readers think your argument is intended to support *both* a belief and an action, you risk confusing them if you in fact support only one, because conceptual and practical claims need different arguments with different kinds of support.

Before readers believe that your answer is relevant to solving a practical problem, they are likely to expect you to support *two* conceptual claims: one claim explains what causes the problem; the other explains how doing something will fix it. But *in addition,* they may also expect you to show the following about your solution:

- It is feasible; it can be implemented in a reasonable time.

- It will cost less to implement than the cost of the problem it solves.

- It will not create a bigger problem than the one it solves.

- It is cheaper or faster than alternative ones—a claim that can be extremely difficult to support.

If readers mistakenly think that you are tacitly proposing a practical claim, they may expect to see those four arguments at least acknowledged. So as you assemble the elements of your argument, be clear about the kind of claim you intend to support: conceptual or practical. If you answer a conceptual question but want to point out its practical applications, build your argument around the answer to the conceptual question and hold off discussing its application until your conclusion, where you can offer it as something worth further consideration (we'll return to this point in chapter 14).

8.2 EVALUATING YOUR CLAIM

We can't tell you how to find your claim or test its truth (other than by testing the argument that supports it). But we can help you roughly evaluate it from the point of view of your readers. They will expect your claim to be both specific and at least potentially significant.

8.2.1 Is Your Claim Specific?

Vague claims lead to vague arguments. The more detailed your claim, the more likely readers will judge it to be substantive, and the more it can help you plan a substantive argument in its support. There are two ways to make it more specific.

SPECIFIC LANGUAGE. Compare these claims:

TV Inflates estimates of crime rates.

The graphic reports of violence on local TV lead regular viewers to overestimate by as much as 150 percent both the rate of crime in their neighborhood and the personal danger to themselves and their families.

The first claim uses only general terms. The second consists of richer, more specific concepts that not only give readers a more specific idea of the claim, but also give the writer a fuller set of concepts to develop in his argument.

Now, we are *not* recommending long, wordy claims for their

own sake. You will benefit if early drafts of your claim have more terms than you ultimately use, but your final claim should be only as specific as your readers need and should include only those concepts that you develop as themes in your argument. But as you assemble the elements of your argument, your first task is to articulate your claim, so at this point, make it as richly explicit as you can. You can fix it later.

SPECIFIC LOGIC. A second kind of specificity depends on how many logical elements your claim includes. Even with its specific language, this claim offers only a single unelaborated proposition:

> Regular TV viewers overestimate by as much as 150 percent both the rate of crime in their neighborhood and the personal danger to themselves and their families.

In the natural and social sciences, claims like this are common, even preferred. But in the humanities, such a claim might seem to be not particularly rich in ideas. For purposes of assembling your argument, try elaborating its logic in two ways:

- Introduce it with a clause beginning with *although* or *even though*.

- Conclude it with a reason-clause beginning with *because*.

For example,

> **Although violent crime is actually decreasing**, regular TV viewers overestimate by as much as 150 percent both the rate of crime in their neighborhood and the personal danger to themselves and their families, **because local TV evening news regularly opens with graphic reports of mayhem and murder in familiar locations, making many believe that crime happens nightly outside their front door.**

While that claim may seem overwritten, it is substantively more explicit. More importantly, it foreshadows three of the five elements that you need for a full argument: *Although I acknowledge X, I claim Y, because of reason Z.*

An introductory *although*-clause can acknowledge alternative views in one of three ways:

- It acknowledges a point of view that conflicts with yours:

    Although most people think they are good judges of the security of their neighborhoods, regular TV viewers overestimate . . .

- It acknowledges a fact that your readers might believe but that your claim qualifies:

    Although violent crime is actually decreasing overall, regular TV viewers overestimate . . .

- It acknowledges a condition that limits the scope or confidence of your claim:

    Although it is difficult to gauge the real feelings about their personal security, regular TV viewers overestimate . . .

If those qualifications are ones that might occur to your readers when they read your claim, then by acknowledging them first, you not only imply that you understand their views, but commit yourself to responding to them in the course of your argument.

On the other hand, a final *because*-clause forecasts reasons for believing the claim—either the most important ones or a general one that encompasses several:

Although many believe that school uniforms help lower the incidence of violence in public schools, the evidence is at best weak, **because no researchers have controlled for other measures that have been instituted at the same time as uniforms** $_{reason\ 1}$ **and because the data reported are statistically suspect.** $_{reason\ 2}$

Again, we do not suggest that in your final draft you offer claims as bloated as our examples. But as you assemble the elements of your argument, the more richly you can articulate a claim, the more comprehensive your argument is likely to be.

8.2.2 Is Your Claim Significant?

After its accuracy, readers will value most highly the significance of your claim, a quality they measure by the degree to which it asks them to change what they think. While you can't precisely quantify it, you can gauge significance by this rough measure: *If readers accept a claim, how many other beliefs must they change?* The most significant claims require an entire research community to change its deepest belief (and that community will resist it accordingly).

Although it is the weakest kind of claim, some research communities will consider a claim significant that asks readers only to accept new information about a subject already studied:

> In what follows, I describe six thirteenth-century grammars of the Welsh language. These grammars have only recently been found and are the only examples of their kind. They help us better appreciate the range of grammars written in the medieval period.

(Recall those reels of newly discovered film, p. 26.)

Readers value research more highly when it offers new knowledge *but also* uses that knowledge to settle what has seemed puzzling, uncertain, inconsistent, or otherwise problematical:

> The relationship between consumer confidence and the stock market has long been debated, but new statistical tools developed in the last few years have shown that there is virtually no relationship whatsoever. . . .

But they value most highly new knowledge that *upsets* what seemed long settled:

> It has long been assumed that the speed of light is constant everywhere at all times, under all conditions, but there is now experimental data suggesting it might not be.

A claim like that will be hotly contested by legions of physicists, because if it is true, they will have to change their minds about lots of things other than the speed of light.

Early in your career, you won't be expected to know what researchers in a field think should be corrected, or at least modified. But you can still estimate the significance of your claim by determining whether readers think it might be worth *contesting*. You can gauge that by judging the apparent significance of its *opposite* claim. For example, consider these two claims:

Shakespeare is a great playwright.

This report summarizes recent research on the disappearance of frogs.

To assess whether either claim is worth contesting, revise it into its opposite: change an affirmative claim into a negative or vice versa:

Shakespeare is *not* a great playwright.

This report does *not* summarize recent research on the disappearance of frogs.

If the reverse of a claim seems self-evidently false (like the first one) or trivial (like the second), then most readers are unlikely to consider the original worth an argument. (It is true, however, that some great thinkers like Copernicus have successfully contradicted apparently self-evident claims such as *Obviously the sun goes around the earth.*)

Especially if you are an advanced researcher, you will measure the significance of your claim by how much it will roil the thinking of your research community. For example, big mammals like the camel and woolly mammoth died out in North America about twelve thousand years ago, either because of disease or because indigenous peoples hunted them to extinction. If you claim they were hunted to death, the many researchers who believe that the earliest Native Americans lived in harmony with nature will have to change their minds about something important to them (and so to that degree, they will resist your claim). But that can be known only by someone in the field aware of those beliefs.

If you are too new to a field to make that assessment, imagine readers like yourself. What did *you* think before you began your own research? How much has your claim changed the way *you* now think? What do *you* understand now that you did not understand before? That's the best way to prepare for reporting research to readers who will ask the same questions. They will put that question most pointedly when they ask the most devastating question any researcher can face: not *Why should I believe that?* but *Why should I care?*

Qualifying Claims to Enhance Your Credibility

Some inexperienced researchers think they are most credible when they are most certain. But flatfooted certainty more often undermines your ethos, and thus your argument. As paradoxical as it may seem, you make a research argument more credible when you acknowledge its limitations. You have already seen that readers expect writers to acknowledge and respond to objections and alternatives (also see chapter 10). When you do, you show that you have dealt with readers openly and honestly; by responding, you show readers why you think their objections do not undermine your argument. But readers look for another kind of limitation as well: you should qualify any claim that is less than entirely certain for all time and in all circumstances.

ACKNOWLEDGE LIMITING CONDITIONS

No claim is free of limiting conditions:

> We can conclude that the epicenter of the earthquake was fifty miles south-southwest of Tokyo, **assuming the instrumentation was accurately calibrated**.

> We believe that aviation manufacturing will not soon match its late-twentieth-century levels, **unless new global conflicts lead to a significant increase in military spending**.

Every claim is subject to countless conditions, so ordinarily you should mention only the ones you expect readers to bring up. Scientists rarely acknowledge that their claims depend on the accuracy of their instruments, because everyone expects them to ensure that they are. But economists often acknowledge limitations on their predictions, both because they depend on circumstances that do change and because readers want to know what conditions to watch for.

Consider mentioning important limiting conditions on your claim even if you think readers would never think of them. (Don't mention more than one or two, and avoid obvious or unlikely conditions.) For example, in this case, not only does the writer show that she was careful, but she also gives a fuller and more accurate picture of the claim:

> Today Franklin D. Roosevelt is revered as one of our most admired historical figures, but toward the end of his second term, he was not popular.*claim* Newspapers, for example, attacked him for promoting socialism, a sign that a modern administration is in trouble. In 1938, 70 percent of Midwest newspapers accused him of wanting the government to manage the banking system. . . . Some have argued otherwise, including Nicholson (1983, 1992) and Wiggins (1973), both of whom offer anecdotal reports that Roosevelt was always in high regard,*acknowledgment* but these reports are supported only by the memories of those who had an interest in deifying FDR.*response* **Unless it can be shown that the newspapers critical of Roosevelt were controlled by special interests,***limitation on claim* their attacks demonstrate significant dissatisfaction with Roosevelt's presidency.*restatement of claim*

USE HEDGES TO LIMIT CERTAINTY

Only rarely can you assert in good conscience that you are 100 percent certain that your evidence is 100 percent reliable and your claims are unqualifiedly true. Careful writers acknowledge these limitations by using modifying words and phrases known as *hedges*. For example, if anyone was ever entitled to be assertive, it was Crick and Watson, the discoverers of the helical structure of DNA. But in the opening of their announcement (condensed), they chose diffidence (the hedges are boldfaced):

> We **wish to suggest a** [note: not *the*] structure for the salt of deoxyribose nucleic acid (D.N.A.). . . . A structure for nucleic acid has already been proposed by Pauling and Corey. . . . **In our opinion**, this structure is unsatisfactory for two reasons:

(1) **We believe** that the material which gives the X-ray diagrams is the salt, not the free acid. . . . (2) **Some** of the van der Waals distances **appear** to be too small.

—J. D. Watson and F. H. C. Crick, "Molecular Structure of Nucleic Acids"

Without the hedges, their claim would be more concise, but also more aggressive. Compare that cautious passage with this more unqualified version of it (most of the more aggressive tone comes from the *absence* of hedges, from the flatfooted lack of any qualification):

We **announce** here **the** structure for the salt of deoxyribose nucleic acid (D.N.A.). . . . A structure for nucleic acid has already been proposed by Pauling and Corey. . . . Their structure **is** unsatisfactory for two reasons: (1) The material which gives their X-ray diagrams is the salt, not the free acid. . . . (2) Their van der Waals distances **are** too small.

When you hedge your language, you give your argument nuance.

Of course, if you hedge too much, you will seem timid or uncertain. But in most fields, readers are not impressed by flatfooted certainty expressed in words like *all, no one, every, always, never,* and so on. Some teachers say they object to all hedging, but what most of them condemn are hedges that qualify every trivial claim. And some fields do tend to use fewer hedges than others. But most careful researchers in most fields know that to seem thoughtfully confident, they must express the limits of that confidence.

Few aspects of your argument affect your ethos more than how you handle its uncertainties and limitations. It takes a deft touch. Hedge too much and you seem mealymouthed; too little, smug. Unfortunately, the line between hedging and fudging is thin. As usual, watch how those in your field manage uncertainty, then do likewise.

CHAPTER NINE

Reasons and Evidence

In this chapter we discuss the two forms of support for a claim: reasons and evidence. We show you how to distinguish between the two, how to use reasons to organize your argument, and how to evaluate the quality of your evidence.

Readers look first for the core of an argument, for its claim and two kinds of support: reasons and evidence. In the sequence of reasons, they see the outline of the logical structure of its support. If they do not see that structure, they are likely to judge your argument shapeless, even incoherent. Evidence, on the other hand, is the bedrock of your argument, the established body of facts that readers need to see before they accept your reasons. If they don't accept your evidence, they are likely to reject your reasons, and with them your claim. So once you know your claim, your next task is to assemble the reasons that support it, and the evidence on which those reasons rest.

9.1 USING REASONS TO PLAN YOUR ARGUMENT

Readers use reasons to decide whether to believe your claim, but they also use them to understand the structure of your report. Reasons outline the logic of your argument, and if each major reason is the point of a section, they outline the report as well. For a complex argument, each reason will be supported with subreasons that serve as the points of subsections of the report.

So as you collect evidence, you can use your reasons (and subreasons) to organize that evidence in a form that anticipates the structure of your report. You can do this as a traditional outline, but at this stage you'll probably find it more helpful to create a chartlike outline known as a "storyboard." Put your main claim and each reason or subreason on its own card (or page). Then put all the evidence that supports an individual reason or subreason on its own card (or page). Finally arrange the cards on a table or wall to make their logical relationships visible, as in the figure below.

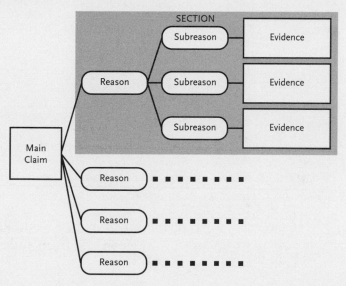

Try out different orders and groupings until you find one that best reflects your current understanding. As your research progresses, try new arrangements. Don't worry about organizing the details; at this point, you want to work with middle-sized chunks that you can arrange in various ways.

If this chart makes your argument look too predictable, don't worry about it. It outlines not your paper but your argument. When you begin to work on a first draft (see chapter 12), you'll have to plan in light of your readers' point of view: how to introduce your problem to make it seem significant to them; how much background to present; and how to order your subclaims; and so on. These are important matters for later, but not now, when you are still discovering what you can make of that mound of notes, summaries, and photocopies.

9.2 THE SLIPPERY DISTINCTION BETWEEN REASONS AND EVIDENCE

On pp. 117–18, we distinguished reasons from evidence. In some contexts the words seem interchangeable:

> You have to base your claim on good reasons.

> You have to base your claim on good evidence.

But they are not synonyms. Compare these two sentences:

> I want to see the evidence that you base your reason on.

> I want to see the reason that you base your evidence on.

That second sentence seems a bit odd because we don't base evidence on reasons; we base reasons on evidence.

• Reasons state why readers should accept a claim. Researchers can think up reasons; they don't think up evidence (or at least they do so at their own risk).

• Evidence is what readers accept as fact, at least for the moment. They think of evidence as "hard" reality, evident to anyone able to observe it.

So when you assemble the elements of your argument, you must start with one or more reasons, but you must base each reason on its own foundation of fact.

The problem is, you don't get to decide whether a statement counts as describing evidence or as just offering another reason—your readers do. If they ask for support for what you offer as evidence, then you have to treat what you thought was evidence as just a reason instead, a reason that you must support with still "harder" evidence. For example, consider this little argument:

American higher education should review its "hands-off" policy toward student drinking off-campus,_{*claim*} because high-risk binge drinking has become a common and dangerous form of behavior._{*reason*} **Injuries and death from it have increased in frequency and intensity, not only at the big "party" schools but among first-year students at small colleges.**_{*evidence*}

In that last sentence, the writer offers what she believes is a "fact" hard enough to treat as evidence. But a skeptical reader might ask, *Are you sure about that? What do you base that on?* In that case, the reader treats that statement not as evidence but as a reason still in need of its own basis in evidence. The writer could add:

Episodes of binge drinking resulting in death or injury by first-year students at colleges with fewer than two thousand students have increased by 19 percent in the last five years.

Of course a *really* skeptical reader could again ask, *Well, how do you know **that's** true?* If so, the writer would have to provide more. If she did her own research, she could produce her raw data and the questionnaires she used to gather them (which themselves are subject to still more skeptical questioning). If she found her data in a source, she could cite it, but then she might be asked to give good reason for accepting it as reliable.

If you can imagine readers asking, *How do you know that? Why should I accept it as a fact?*, then you have not yet hit the bedrock of evidence readers are seeking. And at a time when so-called experts are quick to tell us what to do based on studies we never get to see, experienced readers have learned to view most evidence skeptically. So when you report evidence, be clear about how it was collected and by whom. If it was collected by others, find and cite a source as close to the evidence as you can get.

> ## OUR FOUNDATIONAL CONCEPTION OF EVIDENCE
> When people talk about evidence, they typically use foundational metaphors (as have we): evidence is *hard reality, solid proof,* something we can *see for ourselves.* It's the *bedrock,* the *solid foundation* on which we build arguments. Language like that encourages readers to think of evidence as something independent of their own interpretations and judgments. But data are always constructed and so to some degree shaped by those who collect them—when they decide what to look for, how to record what they see, and how to present what they find. So as you build your argument, try to build it on an unshakable foundation of evidence, but keep in mind that what makes your evidence count as evidence is your readers' willingness to accept it without question, at least for the moment. That way, you may also remember to report it in ways that encourage readers to agree that what you offer is "just the facts."

9.3 EVIDENCE VS. REPORTS OF EVIDENCE

Now a complication: researchers rarely include in any report the actual evidence *itself.* Even if you collect evidence yourself, counting the number of rabbits in a field, in your report you can only *represent* those rabbits in words, numbers, tables, graphs, pictures, recordings, and so on. For example, when a prosecutor says in court, *Jones was dealing drugs, and here is the evidence to prove it,* he can hold up the bag of cocaine, even hand it to jurors so that they can hold in their own hands the "evidence itself." (Of course, both he and the jurors must believe a chemist who says that the white stuff is really cocaine.) But when he *writes* about the case in a law journal, he cannot attach that bag to his article; he can only refer to or describe it.

Unlike prosecutors speaking in a courtroom, researchers almost never share the evidence itself with their readers in their report. The same holds for a researcher who argues this:

> Emotions play a larger role in rational decision-making than most of us think,*claim* because without the help of the emotional centers of the brain, we cannot make rational decisions.*reason*

Persons whose brains have suffered physical damage to their emotional centers cannot make even simple, everyday decisions.*evidence*

That argument doesn't offer as evidence real people with damaged brains unable to make decisions; it can only report observations of their behavior, offer pictures of their brain scans or tables of their reaction times, and so on. (In fact, we much prefer to have researchers report their evidence fairly than for us to have to test brains, read scans, and observe people for ourselves.)

We know the distinction between evidence and reports of evidence may seem like hairsplitting, but it emphasizes two important problems. First, every time you report your own evidence, you change it, usually by cleaning it up and making it more coherent than what you actually saw or counted. Even when you offer seemingly objective quantitative data, you cannot avoid "spinning" them: you must decide what to count, how to categorize the numbers, how to order them. Even photographs and recordings can only represent evidence in a particular way, giving it a slant or shape.

The second problem is that you have to depend on the reports of others, who have already shaped *their* evidence. It is rare for any researcher to rely only on evidence he collected himself, even rarer if he faces a deadline next week. For example, suppose you wanted to back up a claim that the cult of celebrity has distorted rational economic decision-making with evidence of how much more athletes and entertainers earn than do top government officials. You could obtain official reports of government salaries, but those athletes and entertainers would be unlikely to share their check stubs or tax returns (which are themselves reports of reports). So you would have to rely on reports of those reports of salaries. And unless you can talk to the people who did the counting, you'll be four or five reports away from the evidence itself. So as you collect and report evidence, most of it already at least thirdhand, you have to remember that all the reporters in the chain did their own selecting, arranging, and tidying up.

The often dubious quality of reports of reports is why people who read lots of research are so demanding about "proof." If you collected evidence yourself, they want to know what methods you used. If you used sources, they expect you to find primary sources, or if not, sources as close to the evidence itself as you can get. And they want complete citations and a bibliography so that they can go look for themselves. In short, they want to know the complete chain of reports between themselves and the evidence itself. In an age when we are all subjected to research reports and opinion surveys that are at best dubious and at worst faked, you have to give your readers good reason to suspend their justified skepticism, because the last link in that chain of accountability is you.

> ### WHY TRUST REPORTS OF EVIDENCE?
> In the early days of experimental science, researchers conducted experiments before witnesses, reputable scientists who could observe the experiments firsthand and attest to the accuracy of the reported evidence. Contemporary researchers can't rely on witnesses anymore. Instead, each area of study has standardized methodologies for collecting and reporting evidence. Today it is those methodologies that will guarantee that your evidence is reliable. If you follow the procedures for collecting and reporting evidence that have become standard in your field, you encourage readers to accept your evidence at your word, without wanting to see it for themselves or to hear about it from witnesses.

9.4 SELECTING THE RIGHT FORM FOR REPORTING EVIDENCE

You can report evidence in many ways:

- with direct quotations from letters, diaries, books, poems, and so on;

- with words representing objects, images, and events in the form of anecdotes, narratives, and descriptions;

- with photographs, videotapes, films, drawings, and recordings that represent objects and events visually and aurally;

- with tables, graphs, charts, and words representing quantitative data (see chapter 15);

- with summaries and paraphrases of any of the above.

The problem is that different communities of research expect different forms of evidence. Sociologists and economists, for example, prefer data in the form of tables, graphs, and charts. Literary critics rely on quotations from literary texts. Anthropologists and art historians tend to rely not only on verbal descriptions of particular images and events, but also on photographs, videotapes, and sound recordings. Each group accepts other kinds of data, if presented properly, but each is likely to disfavor certain kinds. Literary critics do not expect bar charts to represent the development of an author; most psychologists will be suspicious of mere anecdotes about mental processes.

9.5 RELIABLE EVIDENCE

Once you know the kind of evidence your readers expect, you must test the evidence you have collected by the same criteria that you used to judge your sources (review pp. 76–78): is it sufficient and representative, reported accurately and precisely from an authoritative source? These are not exotic criteria. We all apply them in our most ordinary conversations, even with children. In the following, "P" faults "C" on all those criteria:

C: I need new sneakers.*claim* Look. These seem small.*evidence*

P: Your feet haven't grown that much in a month, and they don't seem to hurt you much [*i.e., I accept that what you offer as evidence could be relevant, but I reject it first because it is not accurate and second because even if it were accurate, "seem small" is not sufficiently precise*].

C: But they're grungy.*reason* Look at this dirt and those raggedy laces.*evidence*

P: Raggedy laces and dirt aren't reason enough to buy new sneakers [*i.e., Your assertion may be factually correct and might be worth considering, but dirt and shoelaces alone are not enough evidence*].

C: Everybody thinks I should get new sneakers.*reason* Harry said so.*evidence*

P: Harry's opinion doesn't matter [*i.e., Even if it's true, other people's opinions are to me not authoritative*].

C: They're hurting me.*reason* Look at how I limp.*evidence*

P: You were walking fine a minute ago [*i.e., Your evidence is not representative*].

If you can imagine yourself as P (or C), you can test the quality of evidence in any research argument, including your own.

Readers judge reports of evidence by P's criteria. They want your evidence to be accurate, sufficient and representative, and precise. And if you didn't gather it, they want it to be from an authoritative source. (Readers may also reject evidence because it is irrelevant or inappropriate, but to apply those criteria, you have to know about warrants, which we discuss again in chapter 11.) So as you assemble the evidence in support of your reasons, screen it before you enter it into your plan.

9.5.1 Report Evidence Accurately

Readers predisposed to be skeptical seize on the smallest flaw in your data, on the most trivial mistake in a quotation or citation, as a sign of your irredeemable unreliability in everything else. If your paper depends on data collected in a lab or in the field, record them completely and clearly, then double-check before, as, and after you write them up. Getting the easy things right shows respect for your readers and is the best training for dealing with the hard things. You can sometimes use even questionable evidence, *if you acknowledge its shaky quality*. In fact, if you point to evidence that seems to support your claim but then reject it as unreliable, you show yourself to be cautious and self-critical—and thus trustworthy.

9.5.2 Provide Sufficient, Representative Evidence

Beginners typically present insufficient evidence. They think they prove a claim when they find support in one quotation, one bit of data, one personal experience (though sometimes only one bit of evidence is sufficient to *reject* a claim).

> Shakespeare must have hated women because those in *Macbeth* are either evil or weak.

Readers usually need more than one bit of data to accept a claim. If your claim is even mildly contestable, find your best evidence, but know that more is always available, and that some of it might be fatal to your claim. Even if you offer lots of evidence, your readers still expect it to be representative of the full range of variation in the available evidence. One Shakespearean play is not representative of all his works, much less of all Elizabethan drama.

9.5.3 Be Appropriately Precise

Your readers also want you to state your evidence precisely. They hear warning bells in certain words that so hedge your claim that they cannot assess its substance:

> The Forest Service has spent **a great deal** of money to prevent forest fires, but there is still **a high probability** of **large, costly** ones.

How much money is *a great deal?* How high is *a high probability*— 30 percent? 50 percent? 80 percent? How many acres are destroyed in a *large* fire? Watch for words like *some, most, many, almost, often, usually, frequently, generally,* and so on. Such words can set appropriate limits of certainty on a claim (see pp. 135–37), but they can also fudge it.

What counts as precise, however, differs from field to field. A physicist measures the life of quarks in infinitesimal fractions of a second, so the tolerable margin of error is vanishingly small. A historian gauging when the Soviet Union was ready to collapse would estimate it in weeks or months. A paleontologist dating a new species might give or take hundreds of thousands of years.

According to the standards of their fields, all three are appropriately precise. (Evidence can also be too precise. A historian would seem foolhardy if she asserted that the Soviet Union reached its point of collapse at 2 P.M. on August 18, 1987.)

Different fields define the criteria for evaluating evidence differently, but each demands that your evidence meet them. If you are a beginner, you will need experience to learn the kinds of evidence readers in your field accept and reject. The painful way to gain that experience is to be the object of their criticism. Less painful is to seek examples of arguments that have failed because their evidence was judged to be unreliable. Listen to lectures and class discussions for the kinds of arguments that your instructors criticize because they think the evidence is weak. Ask for examples of bad arguments. You will better understand what counts as reliable after you see examples of what does not.

Your evidence may be accurate, precise, sufficient, representative, and authoritative, but if readers cannot interpret it quickly, you might as well offer none at all. They will interpret evidence more easily if they understand its relevance to your claim because you added a reason that both supports the claim *and* explains the evidence. Graphically, it looks like this:

For example, what exactly in this table should we see as relevant to the claim in the sentence introducing it?

American consumption of gasoline has contradicted some pessimistic predictions:

	1970	1980	1990	1996
Miles (thousands)	10.3	9.1	10.5	11.3
Consumption (gallons)	830	712	677	698

We need help to interpret the data, to see what we should see, and to know which data are most relevant to the claim. Adding a sentence such as this would help:

American consumption of gasoline has contradicted some pessimistic predictions.*claim* **In 2000 we drove about 23 percent more than we did in 1970, but used 30 percent less fuel.***reason*

	1970	1980	1990	1996
Miles (thousands)	10.3	9.1	10.5	11.3
Consumption (gallons)	830	712	677	698

The added sentence tells us what to look for in the table and how to interpret it. In fact, that sentence does double duty: it not only explains the data, but also offers a reason that supports the claim.

Readers look for the same help when they read a long quotation. Here is a passage that bases a claim about Hamlet directly on the evidence of a quoted passage:

> When Hamlet comes upon his stepfather, Claudius, at prayer, he demonstrates his cool rationality.*claim*
>
>> Now might I do it [kill him] pat, now he is praying:
>> And now I'll do't; and so he goes to heaven;
>> And so am I reveng'd. . . . [Hamlet pauses to think]
>> [But this] villain kills my father; and for that,
>> I, his sole son, do this same villain send to heaven[?]
>> Why, this is hire and salary, not revenge.*evidence*

That argument is not clear. Nothing in the quotation refers to Hamlet's cool reason. In contrast, compare this version:

> When Hamlet comes upon his stepfather, Claudius, at prayer, he demonstrates his cool rationality.*claim* **He impulsively wants to kill Claudius but pauses to reflect. If he kills Claudius while praying, he will send his soul to heaven, but Hamlet wants Claudius damned to hell, so he coldly decides to kill him later:***reason*
>
>> Now might I do it [kill him] pat, now he is praying:
>> And now I'll do't; and so . . . ·*report of evidence*

You can't depend on detailed data or quotations to speak for themselves. Lacking a reason that explains the evidence to them, readers may have to struggle to understand what it means. So always introduce complex evidence with a reason explaining it.

CHAPTER TEN

Acknowledgments and Responses

This chapter discusses matters that can help all researchers, whether beginning or advanced, to convince readers that they are thoughtful and judicious.

As you know by now, the core of your argument is a claim backed by a reason based on evidence. You thicken that core by assembling more reasons, perhaps supporting each with yet more reasons, then laying down a base of evidence on which all those reasons rest. But if you plan your argument only around claims, reasons, and evidence, your readers may think that your argument is flatfooted, even naive. You will seem less like an inquirer amiably engaging intelligent but feisty colleagues in conversation than like a lecturer droning at an empty room.

Since your readers won't be there as you draft your report, you have to *imagine* them asking questions, not just the predictable ones that readers ask about any argument, but ones about yours in particular. It's when you can acknowledge and respond to that imagined questioning, to suggested alternatives and to outright objections, that your report not only speaks in your voice but brings in the voice of others. That's how you most effectively establish a working relationship with readers.

In this chapter we show you how to anticipate two kinds of questions that readers may ask about your argument:

- They may question its intrinsic soundness: the clarity of your claim, the relevance of your reasons, the quality of your evidence.

- They may ask you to consider alternatives—a different way of framing the problem, evidence you haven't considered, warrants that you might not have thought of.

When you acknowledge and respond to both kinds of questions, you construct a written argument that feels like a thoughtful exchange between congenial colleagues.

10.1 QUESTIONING YOUR ARGUMENT

As important as it is to acknowledge other views, don't focus on them as you assemble the core of your own argument. If you do, you may freeze up as you try to imagine every possible alternative. But once you have that core, you have to turn to your imagination to create exceptionally demanding "colleagues" who will help you probe the structure of your argument more forcefully than you hope your readers will.

For this exercise, you might briefly suspend your conception of argument as collaborative inquiry and imagine it not as warfare, but something not unlike a trial. Read your argument as someone would who had a lot at stake in a different solution. That will be hard, because you will know your argument too well, believe in it too much. Fortunately, most of the questions you have to ask are predictable.

First, question your problem as your reader might:

1. Why have you defined the problem in that way? If there is a problem, it involves not what you raise but this other issue.

2. Why do you think there is any problem here at all? I don't see any serious costs if it is not solved. Maybe there is no problem.

3. What kind of problem is this? Is it conceptual or pragmatic? Maybe it should be framed differently.

Now question your solution:

4. Exactly what kind of solution are you proposing? Does your claim ask me to understand something or to do something?

Your solution is conceptual but your problem is practical (or vice versa).

5. Have you stated your claim too strongly? I can think of exceptions and limitations.

6a. Why is your conceptual answer better than others? It doesn't fit in with all this other well-established knowledge.

6b. Why is your practical solution betters than others? I think it will cost too much and create new problems.

If you ask a question that you can't answer, you have to decide whether to find the answer before you go on, or to wait, hoping to find one down the road. Don't be easy with yourself on this one: the best time to fix a problem is when you find it.

Also note where your argument looks weak but is not. If, for example, you anticipate that readers will think your solution has costs that it in fact does not, you can defuse that concern by acknowledging and responding to it:

> It might appear that by recognizing genetic factors in homosexuality we challenge the relevance of free will to sexual orientation. But in fact . . .

Next, question your support, focusing first on your evidence. Some objections probe your evidence:

1. I'd like to see a different kind of evidence. We need hard numbers, not anecdotes. (Or, we want to hear about real people, not cold numbers.)

2. It isn't accurate. The numbers don't add up.

3. It isn't precise enough. What do you mean by "many"?

4. It isn't current. There is more recent research on this.

5. It isn't representative. You didn't get data from . . .

6. It isn't authoritative. Smith is no expert on this matter.

The toughest objection, however, is usually this one:

7. You need more evidence. One quotation does not establish a pattern.

Almost all researchers have difficulty finding enough of the best evidence to make a solid case, a problem even worse for those working on short deadlines. Teachers predictably grumble when students think that any evidence they find is enough. But the problem is even worse when readers resist a solution because they have a lot at stake in what they believe; when they do, you can expect them to demand more evidence of higher quality, perhaps more than you have time to find. So if you feel your evidence is less than unassailable, you may want to make a note to admit that candidly before your reader rejects your argument because you didn't.

Finally, readers may also feel that your claim just does not follow from your reasons or even that a reason is irrelevant to your claim. But that is an issue so vexed that we devote all of chapter 11 to it.

In sum: A crucial step in assembling your argument is to test it as your readers will, even in ways they might not, and then to acknowledge and respond to at least the most important objections that you can imagine them raising. If you don't show readers that you have put your own argument through the wringer, they will do it for you.

10.2 FINDING ALTERNATIVES TO YOUR ARGUMENT

When you acknowledge weaknesses in your argument, you add to your credibility by showing readers that you are not just making an honest case but dealing fairly with them. But that is a defensive move, not one that actively brings their views into your argument. For that, you have to imagine those views and how they point to possible alternatives. That is easiest when you know your readers well, but even if you have to guess about them, you can rely on some strategies to help you do that. (At this point, return to the amiable, collaborative image of readers.)

10.2.1 Alternatives in Your Sources

As you read your sources, you will find examples of how others have thought, if not about your specific question, then at least about your topic. Note where a source takes an approach different from yours, focuses on different aspects of the problem, and so on. Especially note outright disagreement: even if the source does not help you make your argument, it can help you see alternatives to it. Note any alternative positions the source cites; you may want to acknowledge them as well. Finally, when you have finished taking notes, reflect on how that writer sees your issue differently from you. You may even want to sketch out some of those alternatives in your summary.

Don't ignore evidence because you decide that it is unreliable or irrelevant. If you think readers might consider it relevant, you may want to acknowledge it but respond by explaining why you didn't use it. That's one way to compensate for not having enough evidence of your own. If, as you take notes, you pay as much attention to disagreements and alternatives to your claim as you do to data that support it, you'll not only understand your problem better, but you'll better anticipate weaknesses or limits to your argument that may be decisive for readers.

10.2.2 Three Predictable Alternatives

When you look for alternatives in your sources, you have to look out for almost anything—alternative evidence, interpretations of evidence, judgments of the reliability of evidence, conclusions, lines of reasoning, and so on. But there are three kinds of alternatives that at least some readers are likely to think of.

I. **"But there are causes in addition to the one you claim."** If your argument concerns cause-and-effect, remember that no effect has a single cause and no cause has a single effect. If you argue that X causes Y, every reader will think of countless other factors without which Y can't happen. The Soviet Union may have collapsed partly because President Reagan's military buildup forced it to spend more on arms than its economy could afford. But an informed reader could list many other factors, ranging from decades of poor economic performance to political corruption to self-destructive ideology. So if you focus on one cause out of many, acknowledge the others, and if you feel readers may think that some cause deserves more attention than you give it, acknowledge that view and explain why you focus on your own.

2. "But what about these counterexamples?" No matter how copious your evidence, readers are likely to think of counterexamples that they think undermine your argument. So you have to think of them first, then acknowledge the more plausible ones, especially if they are vivid. Then explain why you don't consider them as damaging as your reader might. Be particularly wary when you make claims about a phenomenon with a wide range of variation, such as climate data. Readers who do not understand statistical reasoning will focus on an aberrant case, even though it falls within a normal distribution: a cold Fourth of July does not disprove a claim about global warming, any more than a warm Christmas proves it.

3. "I don't define X as you do. To me, X means . . ." To accept your claim, readers must accept your definitions. If you are researching nicotine addiction, your readers must understand what you mean by *addiction*. Does it mean just a strong craving, a craving that some people can't resist, or a craving that *no one* can resist? You can find definitions ranging from a few lines in a dictionary to pages in a medical reference work. But regardless of what those sources say, readers will try to define your terms to suit their views. Cigarette manufacturers long argued that cigarettes are not addictive because some people can quit; their critics argued that cigarettes are addictive because others can't.

When your argument turns on the meaning of a disputed term, define it in a way that supports your solution, then make an argument supporting your definition. (Don't begin: *According to* Webster's, addiction *means . . .*) Be aware of plausible alternative definitions you may need to acknowledge. If you use a technical term that also has a more common meaning that your readers use (like *social class* or *theory*), acknowledge the ordinary definition and explain why you need the technical one to solve your problem. If you do not use a technical definition as expert readers would expect you to, acknowledge that and explain why you use the more common meaning.

10.3 DECIDING WHAT TO ACKNOWLEDGE

If you can imagine just a few of the questions, alternatives, and objections your readers might have, you'll face a Goldilocks moment: acknowledge too many and you distract readers from the core of your argument; acknowledge too few and you seem indifferent to or even ignorant of your readers. You need to figure out how many will feel "just right."

10.3.1 Selecting Alternatives to Respond To

To narrow your list of alternatives or objections, consider these priorities:

- plausible charges of apparent weaknesses that you can rebut;

- alternative lines of argument that have been important in your field;

- alternative conclusions that readers *want* to be true;

- alternative evidence that readers know;

- important counterexamples that you have to explain away.

Next, look for alternatives that let you repeat a part of your argument. For example, if readers might think of exceptions to your definition that in fact are not, acknowledge them and use the response to reinforce your definition:

Some have argued that food can be addictive, but remember that we are here concerned only with substances for which addiction is the norm, not those . . .

Or if readers might think of an alternative solution close to yours, use it to reiterate the virtues of your solution:

> Given the cost of bringing in tenure-line faculty for a Center for English Language, we could start slowly by hiring part-time adjuncts with experience as ESL tutors. Tutors would certainly improve the situation, but only for a short time. If we put a Band-Aid on the problem now, we are unlikely to be able to generate momentum for a more permanent solution later. . . .

Finally, acknowledge alternatives that may particularly appeal to your readers, but only if you can respond without seeming to be dismissive. Better to ignore something your readers like than to disparage it.

10.3.2 Acknowledging Questions You Can't Answer

All researchers fear questions they can't answer. If you discover a flaw that you cannot fix or explain away, you can try to redefine your problem or rebuild your argument to avoid it. But if you cannot, you face a tough decision. You could ignore the problem, hoping readers won't notice. But that's dishonest. And if they do notice, you have a bigger problem because they will doubt your competence, and if they think you tried to hide it, they will question your honesty. In either case, the damage could be fatal, not only to your argument but to your ethos and reputation.

Our advice may seem naive, but it is useful: Openly acknowledge the problem and respond that

- the rest of your argument more than balances the flaw;

- while the flaw is serious, further research would show a way around it;

- while the flaw makes it impossible to accept your claim fully, your argument offers important insight into the question and suggests what a better answer would need.

Occasionally researchers turn failure into success by turning the claim they *wanted* to support but couldn't into a hypothesis that people might *think* is a reasonable solution to a problem but turns out not to be. Then they go on to show why not:

> It might seem that when jurors hear the facts of a case in a form that focuses on the victim and emphasizes her suffering, they will be more willing to blame the accused. That is, after all, the standard practice of the best plaintiff's lawyers. But in fact, we found no correlation between . . .

Experienced researchers and teachers understand that truth is always complicated, usually ambiguous, always contestable. They will think better of your argument and of you if you acknowledge its limits, especially those that squeeze you more than you like. Concessions invite readers into the conversation by legitimizing their views.

10.4 RESPONSES AS SUBORDINATE ARGUMENTS

You have to respond to your imagined readers' most thoughtful alternatives and objections with an argument. Even the most minimal response gives a reason for accepting, limiting, or rejecting what you have acknowledged:

> Some have argued that food can be addictive,*acknowledgment/claim to be rejected* but remember that we are here concerned only with substances for which addiction is the norm.*reason for rejecting claim* Some people who taste chocolate once may be unable to resist it thereafter, but the number who crave chocolate is a fraction of those who after trying crack cocaine just once are immediately addicted to it.*reason* Chernowitz (1998) found that just one exposure to crack cocaine resulted . . . *report of evidence*

For more substantial responses, you need a full argument, with multiple reasons, evidence, and perhaps even warrants and additional acknowledgments and responses. (At this point, just add acknowledgments and responses to the appropriate places in the working plan of your argument. In chapter 12 we'll discuss where to put them in the plan of your first draft.)

When you respond to alternatives with reasons and evidence

for rejecting them, you "thicken" your argument, making it increasingly rich and complex, enhancing your credibility as someone not given to oversimplifying complex issues. Readers will respect you and your argument when you bring their voices into your report by acknowledging their alternatives and objections. But as we said, this is a Goldilocks choice: not too much, not too little. Only experience tells you what's just right. So notice how experts in your field do it, and ask readers for advice and criticism.

 *The Vocabulary of*
Acknowledgment and Response

Some writers fail to acknowledge alternatives because they can't think of anything to acknowledge. The strategies in this chapter will help overcome that problem. Others can think of them but fear that if they acknowledge them, they weaken their argument. In fact, they raise its credibility in the eyes of most readers. A third reason writers don't acknowledge objections and alternatives is the easiest to fix: they can imagine objections but lack the vocabulary to express them. What follows is that vocabulary. To be sure (that's one of those terms right there), your first efforts in using these words and phrases may seem awkward (*may* is common in acknowledgments), *but* (a response typically begins with *but* or *however*) as you use them, they will soon feel natural.

ACKNOWLEDGING
When you respond to an alternative or objection, you can mention and dismiss it or address it at length. We offer these expressions roughly in that order, from most dismissive to most respectful. (Brackets and slashes indicate alternative choices.)

1. You can dismiss an objection or alternative by introducing it with *despite, regardless of,* or *notwithstanding:*

   [Despite/Regardless of/Notwithstanding] Congress's claims that it wants to cut taxes,*acknowledgment* the public believes that . . . *response*

   Use *although, while,* and *even though* in the same way:

   [Although/While/Even though] there are economic problems in Hong Kong,*acknowledgment* Southeast Asia remains a strong . . . *response*

2. You can signal an acknowledgment indirectly with *seem, appear, may,* and *could,* or with an adverb like *plausibly, justifiably, reasonably, surprisingly,* or even *certainly:*

In his letters, Lincoln expresses what [**seems/appears**] to be depression.*acknowledgment* But those who observed him . . .*response*

This proposal [**may have/plausibly has**] some merit,*acknowledgment* but we . . .*response*

3. You can acknowledge alternatives by attributing them to an unnamed source or to no source at all, which gives a little weight to the objection:

It is easy to [**think/imagine/say/claim/argue**] that taxes should . . .

There is [**another/alternative/possible**] [**explanation/line of argument/account/possibility**].

Some evidence [**might/may/can/could/does**] [**suggest/indicate/point to/lead some to think**] that we should . . .

4. You can attribute an alternative to a more specific source, giving it more weight:

There are [**some/many/a few**] who [**might/may/could/would**] [**say/think/argue/claim/charge/object**] that Cuba is not . . .

Note that researchers sometimes weaken their case by prematurely downgrading those they will disagree with:

Some **naive** researchers have claimed that . . .

The **occasionally careless** historian H has even claimed that . . .

It's usually best to save your criticism for the response, and to direct it at the work rather than the person.

5. You can acknowledge an alternative in your own voice, with a passive verb or with an adverb such as *admittedly, granted, to be sure,* and so on, conceding it some validity:

I [understand/know/realize] that liberals believe in . . . ,
but . . .

It is [true/possible/likely/certain] that no good evidence proves
that coffee causes cancer. . . . However, . . .

It [must/should/can] be [admitted/acknowledged/noted/con-
ceded] that no good evidence proves that . . . Nevertheless, . . .

[Granted/Admittedly/True/To be sure/Certainly/Of course],
Adams has claimed . . . However, . . .

We [could/can/might/may/would] [say/argue/claim/think] that
spending on the arts supports pornographic . . .

RESPONDING
Begin your response with contradicting language like *but, however,* or *on the other hand.* After you state your response, offer
some support for it, because that response is a claim. You can
respond in ways that range from tactful to blunt.

1. You can regret that *you* don't entirely understand:

   But [I do not quite understand how . . ./I find it difficult to see
   how . . ./It is not clear to me how] X can claim that, when . . .

2. Or you can note that there are unsettled issues:

   But there are other issues here . . ./There remains the problem
   of . . .

3. You can respond more bluntly, claiming the acknowledged
   position is irrelevant or unreliable:

   But as insightful as that may be, it [ignores/is irrelevant to/does
   not bear on] the issue at hand.

   But the [evidence/reasoning] is [unreliable/shaky/thin].

   But the argument is [untenable/weak/confused/simplistic].

   But the argument [overlooks/ignores/misses] key factors. . . .

You have to decide just how forceful your blunt rejection should be; if the alternative seems to you obviously flawed, say so, but focus on the work, not the person.

ADDRESSING LOGICAL ERROR

When you think a writer might not have thought through an issue carefully, you usually should say so civilly. Here are a few possibilities:

> That evidence is important, but we must look at all the available evidence.

> That explains some of the problem, but it is too complex for a single explanation.

> That principle holds in many cases, but we must also consider the cases it overlooks.

CHAPTER ELEVEN

Warrants

This chapter raises an issue more complex than you may want to face, especially if you are just beginning: the logical relevance of your reasons to your claims. In the long run, however, every researcher should work to understand it.

Researchers owe their readers their best reasons, backed with more than enough of the best available evidence. But even if readers accept your reasons as true, they may still not accept your claim if they think your reasons are *irrelevant* to it. We explain and demonstrate the relevance of a reason to a claim with the fifth element of argument—a warrant.

A warrant is sometimes called a *commonplace,* a common-sense generalization about the world that everyone considers self-evident: *Where there's smoke, there's fire.* But some warrants are so specific to a particular community that they virtually define its special habits of mind: *When different species share little DNA, we can conclude that they diverged earlier than species that share more DNA.* Like all commonplaces and habits of mind, we sometimes make them explicit, but more often we take them for granted.

In this chapter we show how warrants explain your reasoning, how to know when you must state them, and how to formulate and test them. But first a caution: Warrants are the most abstract, difficult element in an argument to understand and manage. Everyone struggles to grasp them, and rhetorical theorists debate them. So if at the end of this chapter you still have questions, you are in good company (including, from time to time, the three of us).

11.1 HOW WARRANTS WORK

Suppose your friend makes this argument:

> Despite Congress's doubling the budget to reduce drug smuggling, the amount of drugs smuggled into this country has risen._{reason} Clearly, we are wasting our money._{claim}

You respond:

> Why should the fact that smuggling has increased despite a bigger budget to prevent it mean that we are wasting money? I don't see how that follows.

To persuade you to accept that reason as supporting that claim, your friend would have to respond with a general principle that explains why it does. His principle would consist of two parts, a general circumstance and a general consequence that reliably follows from it:

> When more resources are invested to prevent something but its incidence goes up,_{general circumstance} those resources have been wasted._{general consequence}

If you accept the general principle (you might not), then you should accept the same relationship between any *specific instance* of that circumstance and any *specific instance* of that consequence. If you accept that the general consequence follows from the general circumstance, then you should also accept that the specific consequence follows from the specific circumstance.

We can represent how a warrant "covers" a reason and claim graphically like this:

General circumstance	*predictably leads to*	General consequence
When more resources are invested to prevent something but its incidence goes up,		those resources have been wasted.
✔		✔
Despite Congress's doubling the budget to reduce drug smuggling, the amount of drugs smuggled into this country has risen.*reason*	***therefore***	We are wasting our money.*claim*
Specific circumstance	*lets us infer*	Specific consequence

The check marks indicate that we think

- the specific circumstance (*Despite Congress's doubling the budget to reduce drug smuggling, the amount of drugs smuggled into this country has risen*) is a good instance of the general circumstance (*more resources are invested to prevent something but its incidence goes up*);

- the specific consequence (*We are wasting our money*) is a good instance of the general consequence (*resources have been wasted*).

If the warrant and reason are true and the reason and claim are good instances of the warrant, then the claim must be true. Of course, the warrant will not "work" if you don't accept it as a true general principle. In that case, your friend either has to make a case to convince you to accept it or find another applicable one that you do accept.

Writers usually offer warrants to connect a reason and a claim, so that's what we will concentrate on here. But you should know that you can also offer warrants to explain how evidence is relevant to a reason. Since reasons are (sub)claims, warrants connect a reason to its supporting evidence just as they connect a claim to its supporting reason.

11.2 WHAT WARRANTS LOOK LIKE

In practice, writers state warrants in many ways, from direct to oblique:

> If a problem continues, resources invested in prevention are wasted.

> Spending money for nothing is a waste.

> An ounce of prevention is wasted if you still need the cure.

But however it is stated, a warrant always has those two parts: a general circumstance and the general consequence that readers should infer. The parts can relate by cause-and-effect (*Rain causes wet streets*), one-thing-is-the-sign-of-another (*Cold hands, warm heart*), a rule of behavior (*Look both ways before you cross the street*), a definition (*A three-sided figure is a triangle*), a principle of reasoning (*Sufficient representative data are necessary for any reliable generalization*), or by any other principle that links a condition and a consequence.

But for our purpose here, this next way of stating warrants is most useful because it clearly distinguishes the two parts that every warrant must have:

> When(ever) X, then Y.

This formulation helps you test the connection between a specific condition and a specific consequence. You can then restate the warrant however you like.

11.3 KNOWING WHEN TO STATE A WARRANT

Research reports involve countless principles of reasoning, most of them so deeply embedded in our assumptions and tacit knowledge that we would never question them. That's why researchers state warrants only when they think their readers will question the relevance of a reason to a claim. Look especially for the following three cases:

- You can assume that some readers will have questions if you use a principle of reasoning that you know is new or controversial in your field.

In that case, explicitly state it as a warrant; then justify it, preferably by referring to authoritative figures who also use and defend the principle. You are not likely to convince those already set against it, but you will at least acknowledge that you know your position is controversial and show that you are not alone in holding it.

- Readers will also look for warrants if they are unfamiliar with the kind of argument you are making.

If you are writing as a specialist in a field to readers who are not, find places where you use reasons that only specialists would use. If the principle behind that reason is one only specialists would recognize, explain it with a warrant. If readers are generally familiar with your kind of argument, look for places where you reason in surprising or unconventional ways. Even if readers recognize an unconventional principle of reasoning, you can diffuse some of their resistance by explicitly stating *and defending* the warrant that explains it.

- Readers are more likely to question your reasoning when they resist your claim because they just don't want it to be true.

In that case, start with a warrant that you think they will accept before you lay out the reason and claim you expect them to resist. They may not like the claim any better, but you will at least force them to see that their resistance is illogical. For example, consider this argument:

Homosexuality must have a strong genetic component $_{claim}$ because so many of its characteristics appear in the feelings and behavior of children who have no contact with homosexuals but become homosexual adults.*reason*

Some readers resist that claim because they believe that sexual orientation is a matter of free will and that any genetic basis for homosexuality would compromise their moral objections to it. A writer might not be able to overcome their strongly held beliefs, but if he had good evidence to support that reason, he might get them to consider the claim if he first convinced them to accept a warrant connecting that reason and claim:

When children manifest behavior arising not from teaching or modeling, but spontaneously, that behavior is genetically based.*warrant* Homosexuality must therefore have a strong genetic component*claim* because . . .*reason*

If readers think that both warrant and reason are true, and that the specific reason and claim are good examples of the warrant, they are logically obliged to accept the claim. If they do not, you know that no rational argument is likely to change their minds.

WHAT YOU DON'T SAY SAYS WHO YOU ARE

You show consideration for readers when you offer warrants to explain principles in your field that they may not recognize. But it is an equally strong gesture when you keep silent about warrants you could have stated. Warrants articulate the principles of reasoning that form the intellectual fabric of a research community. So when you are silent about warrants exclusive to your field, you exclude readers not in the know and implicitly claim that you are a knowledgeable insider. One way or the other, warrants significantly affect how readers perceive your ethos.

11.4 TESTING YOUR WARRANTS

Assume that your readers are most likely to challenge your warrants when they strongly resist your claims. Consider this little argument:

We believe that, contrary to popular belief, gun ownership was not widespread in the first half of the nineteenth century in America or before,*claim* because guns were so rarely mentioned in wills.*reason* A review of 4,465 wills filed in seven states from 1750–1850 shows that only 11 percent of them mention a long gun or handgun. . . .*report of evidence*

You can expect that claim to be resisted by those whose image of America includes widespread gun ownership stretching back before the Revolution. Even if they accept that guns were rarely mentioned in wills, they may still object: *Why should the fact that guns were rarely mentioned in wills count as a reason for believing that few people owned one?*

If a writer anticipated that objection, she could begin with a warrant:

> In the eighteenth and nineteenth centuries, most household objects were regularly listed in wills, especially if they were valuable objects like guns. So when someone failed to mention such an object, he probably did not own one.

But the moment she states that warrant, she should ask herself three questions:

- Is that warrant true and appropriately limited?

- Does it apply to the reason and claim?

- Is it appropriate and persuasive for the readers of this argument?

11.4.1 Is Your Warrant True and Appropriately Limited?

If your readers think your warrants are just false, no amount of reasons and evidence can save your claim.

> Nonhuman creatures are mere biological objects without any inner life and so should not be objects of pity or concern.*warrant* Since apes used in medical experiments experience nothing like

human emotions or feelings,*reason* we should not waste money trying to make their conditions more comfortable.*claim*

Half a century ago, most psychologists believed that warrant to be true. Almost none do today.

A warrant can be basically true but stated too generally. For example, here is that warrant about gun ownership with no qualifications or hedges:

> In the eighteenth and nineteenth centuries, household objects were listed in wills.

That's too strong. Scaled back, it might be more acceptable:

> In the eighteenth and nineteenth centuries, household objects **considered valuable by their owners** were **usually** listed in wills.

These tests also apply to *creating* a warrant when you need one. A good principle is to create a warrant that is only a bit more general than the reason and claim, and that does not depend on words like *everyone, any, never,* and *always.* It is particularly challenging to formulate a warrant when the reason and claim are already general statements. When that's the case, the warrant has to be more general yet. For example:

> Belief in astrology resists logical argument*claim* because people tend to remember vivid coincidences between a prediction and a random daily event better than they remember the many more times a prediction failed.*reason*

We can find a warrant for that by restating the specific reason and claim in the *When X, then Y* form:

> When people remember vivid coincidences between an astrological prediction and a random daily event better than they remember the many more times when a prediction failed,*reason side* their belief in astrology resists logical argument.*claim side*

Then we revise both sides to make them more general:

When people generalize on one vivid coincidence,*reason side* they
do not think logically.*claim side*

That warrant is, in fact, an important principle of decision science, but is it always true, in all circumstances, at all times? If
not, we open it to exceptions that may lead readers to reject not
only the warrant but the whole argument.

11.4.2 **Does Your Warrant Actually Apply to Your Reason and Claim?**
This test for warrants addresses a matter that has vexed logicians
and rhetoricians for more than two thousand years: How does a
warrant connect a reason to a claim *validly?* When your reasons
and evidence are untrue, you can correct them; when they are
unclear, you can clarify them. But when someone says your claim
is *unwarranted,* or refers to it by the Latin term *non sequitur* ("it
doesn't follow"), you have to analyze the logic of your argument.
Here is a simple example:

Alex: You should buy a gun, because you live alone.

Anya: Why should my living alone mean I should buy a gun?

Alex: Whenever you live in insecure circumstances, you should protect
yourself.

Anya: But living alone does not mean that my life is insecure.

Anya complains that Alex's reason is not a good instance of the
reason side of his warrant, at least for her, because living alone
is not an instance of being insecure.

But testing other arguments can be harder. Here, for example,
is a subtly flawed argument about the effect of TV violence on
children (we should alert you that what follows requires close
attention):

Few doubt that when we expose children to examples of courage
and generosity, we influence them for the better. How can we
then deny that when they are constantly exposed to images of sadistic violence, they are influenced for the worse?*warrant* Data
show that violence among children 12–16 is rising faster than

among any other age group.*reason* Brown (1997) has shown that
. . .*evidence* We can no longer ignore the conclusion that TV violence, even in cartoons, is a destructive influence on our children today.*claim*

To diagnose what is wrong here, we break the warrant into its
two parts, and then align the reason and claim under them.

General circumstance	*predictably leads to*	General consequence
When children are constantly exposed to images of sadistic violence,		they are influenced for the worse.
?		**?**
Data show that violence among children 12–16 is rising faster than among any other age group.*reason*	*therefore*	TV violence is a destructive influence on our children today.*claim*
Specific circumstance	*lets us infer*	**Specific consequence**

Now we see that the specific circumstance is not a good instance
of the general one: *rising violence* is not an instance of *children
being exposed to images of violence*. Similarly, the specific consequence is not a good instance of the general one: *TV violence is
destructive* is not an instance of *children being influenced for the
worse*, because it is too specific. So even if all of those statements
are true (arguably they are), they do not add up to a valid argument,
because the warrant covers neither the reason nor the claim.

To fix that argument, we would have to revise both the reason
and claim to fit the warrant (or the warrant to fit the reason and
claim):

Few doubt that when we expose children to examples of courage
and generosity, we influence them for the better. How can we
then deny that when they are constantly exposed to images of sadistic violence, they are influenced for the worse? All our data

show that violence among children 12–16 is rising faster than among any other age group. This violence results from many factors, but we can no longer ignore the conclusion that be-cause **television is the major source of children's images of vio-lence,**$_{reason}$ **they are becoming violent because of it.**$_{claim}$

The evidence and claim seem closer to the kind that the warrant admits:

General circumstance	*predictably leads to*	General consequence
When children are con-stantly exposed to im-ages of sadistic violence,		they are influenced by those images for the worse.
✔		✔
Television is the major source of children's im-ages of violence.$_{reason}$	***therefore***	Children are becom-ing violent because of it.$_{claim}$
Specific circumstance	*lets us infer*	Specific consequence

But a reader keen to derail the argument might still object:

Hang on. Your reason does not, in fact, fit your warrant. It is true—images of violence do appear on television. But I don't be-lieve that those images are "sadistic." A lot of it is cartoon vio-lence. Therefore, your warrant cannot cover your reason because your reason is not a good instance of your warrant. Furthermore, your claim—"becoming violent"—is more extreme than "influ-ence for the worse." It is too specific and so goes beyond the claim your warrant allows.

As we said, this is not easy stuff.

11.4.3 Is Your Warrant Appropriate to Your Readers' Research Community?

Law students get a painful lesson in learning to make legal argu-ments when they find out that many commonsense warrants that most of us believe have no place in their world of legal reasoning. For example, like most of us, they start law school holding this commonsense belief that we can express as a warrant:

> When someone does another an injustice, our legal institutions should correct it.

But law students have to unlearn such commonsense warrants, because other warrants may trump them. For example,

> When you fail to meet legal obligations, even inadvertently, you must suffer the consequences.

More specifically,

> When old people forget to pay real estate taxes, others can buy their house for back taxes and evict them.

Against their most decent instincts, law students have to learn that justice is not what most of us want it to be, but what courts say it is.

Warrants help you understand why important issues are so endlessly contestable: why, when you feel you have a watertight case, your readers still say, *Wait a minute. What about . . . ? I don't agree that your evidence counts as . . .*

Even more troublesome, readers may offer competing warrants:

> When unions want to express their political views, they have a constitutionally protected right to do so. The local teachers union believes real estate taxes should be raised, so they have a right to picket the school board meeting.

> When there is no unanimous agreement in a group, the group should not express a controversial opinion. Not every member of the local teachers union thinks real estate taxes should be raised, so it should not picket the school board meeting.

What reasons and evidence could we offer to prove either warrant? And what higher-order warrants would cover those reasons?

Of all our disagreements with one another, those involving warrants cut the deepest.

11.5 CHALLENGING THE WARRANTS OF OTHERS

If it is hard to convince readers to accept a new warrant, it is more difficult to get them to give one up they believe. If you want to build your argument on warrants that challenge your readers' basic principles, start by imagining how readers would *defend* the warrant you want to challenge. For example, an economist might argue:

> The population of Zackland must be controlled$_{claim}$ because it is outstripping its resources and heading for disaster.$_{reason}$ **When a population grows beyond its resources, only a reduction in population will save the country from collapse.**$_{warrant}$

If someone challenged that warrant, he might back it with economic analysis:

> When countries A, B, and C exceeded their means, each collapsed. They tried to prevent collapse by every means other than population control, but it did no good.$_{reason}$ **When societies reach a point where their population exceeds their resources, the only way they can prevent collapse is to reduce their population.**$_{claim/warrant}$

But a religious person might challenge that argument with another claim based on a warrant grounded not in economic principles but moral ones:

> It doesn't make any difference what the economic consequences might be; it is immoral to discourage married couples from having children.$_{claim}$ **When people are advised to defy God's will as revealed in our holy books, that advice is sinful.**$_{warrant}$

A third person might also reject population control but offer yet a different warrant:

Whenever we put our minds to a problem of limited resources, we can solve it.

Asked what backs up such a warrant, that third person might say, *Well, I believe in a can-do attitude. It's the American way.* This last warrant is based not on data or religious belief but on cultural conditioning. Those three different warrants are each supported in different ways: by economic data, by a system of revealed truth, by cultural inheritance. To challenge them, you have to challenge their support, each in its own way.

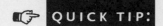 *Some Strategies for Challenging Warrants*

Since warrants can be based on fundamentally different princi-ples of reasoning, you have to challenge them in different ways.

WARRANTS BASED ON EXPERIENCE

Asked to defend a warrant based on experience, we refer to every-day experience or to reliable reports by others.

Where there's smoke, there's fire.

When certain insecticides leach into the ecosystem, eggshells of wild birds become so weak that fewer chicks hatch and the bird population falls.

CHALLENGES: To challenge those warrants, you have three choices, all difficult: (1) find counterexamples that cannot be dis-missed as special cases; (2) challenge the reliability of their expe-rience; or (3) argue that the evidence is not relevant to the war-rant. Choose the first strategy if you have good counterexamples. You can argue without directly discounting the experience or the reasoning of your readers. For the other two, you have to tackle readers head-on.

WARRANTS BASED ON AUTHORITY

We believe some people because of their expertise, position, or charisma.

When authority X says Y, Y must be so.

CHALLENGES: Challenging authority is difficult. The easiest— and friendliest—way is to argue that, on this matter, the authority does not have all the information or has reached beyond her core area of expertise. The most direct way is to give good reason not to take her at her word, because she is no authority at all.

WARRANTS BASED ON SYSTEMS OF KNOWLEDGE AND BELIEF

These warrants are backed by systems of definitions, principles, or theories:

From mathematics: When we add two odd numbers, we get an even one.

From religion: When we commit adultery, we commit a sin.

From law: When we drive without a license, we commit a misdemeanor.

CHALLENGES: When you challenge these warrants, "facts" are largely irrelevant. You must either challenge the system, always difficult, or show that the case does not fall under the warrant: what about driving in my own driveway?

GENERAL CULTURAL WARRANTS

These are the warrants that seem just "common sense" to members of a particular culture. Some are backed by empirical experience, but many are not:

Early to bed, early to rise, makes you healthy, wealthy, and wise.

Whenever a king wants to abuse his subjects, he may.

It is always wrong to mock someone from another culture.

CHALLENGES: Warrants like these change over time, but slowly. You can challenge them, but readers will resist your attempts to change them because you will seem to be challenging their culture.

METHODOLOGICAL WARRANTS

Think of these as "meta-warrants," general patterns of thought that have no content until applied to specific cases. We use them to explain our reasoning:

Generalization: When many cases of X have the quality Y, then X is characterized by Y.

Analogy: When X is like Y in certain respects, then X will be like Y in other respects.

Cause-effect: When Y occurs if and only if X occurs first, then X may cause Y.

Sign: When Y regularly occurs before, during, or after X, Y is a sign of X.

CHALLENGES: Philosophers have questioned even these warrants, but in matters of practical argumentation, we challenge only their application or point out limiting conditions: *Yes, we can analogize X to Y, but not if . . .*

WARRANTS BASED ON ARTICLES OF FAITH

Some warrants are beyond challenge: Jefferson invoked that kind of warrant when he wrote, "We hold these truths to be self-evident. . . ."

When a claim is directly experienced as revealed truth, that claim is true.

When a claim is in accordance with divine teachings, it must be true.

Such warrants are backed not by any confirmable evidence but simply by the believers' inner certainty. They are statements of faith, requiring no argument, no evidence.

CHALLENGES: It makes no sense to challenge these warrants, because no argument could support or undermine them. The best you can do is offer an equally unargued alternative. If you encounter them as you gather your data, either ignore them or decide to study them from an entirely different perspective: not as a subject for research but as an inquiry about the meaning of life.

PART IV

Preparing to Draft, Drafting, and Revising

Prologue

No formula can tell you when to start drafting. Booth begins, in a sense, "too early"; then once his ideas become clearer, he has to face up to the nasty task of deciding what to throw out. Colomb is an inveterate outliner, producing a dozen outlines and two or three "advance summaries." Williams tries out as many versions as do Booth and Colomb, but in his head; he writes as he goes but starts a serious draft only when he has a sense of the whole. You, too, must find your own best way to get started.

You can prepare for that moment if you have been writing summaries, analyses, and critiques from the start. You *know* you are ready when you can do this:

- you have figured out what your *readers* are like, why they should care about your problem; and the kind of *character* you want to project;

- you can sketch the *question* that defines the gap in knowledge or understanding that you want to resolve and an *answer* in a sentence or two;

- you can sketch the *support* for your claim: your main reasons and evidence;

- you have an idea of the kinds of questions, alternatives, and objections your readers are likely to raise, and you can respond to them;

- you know the major warrants that you must state and perhaps support.

When experienced writers think they are ready to start a draft, though, they know they won't march straight to a finished product. They know they will go down blind alleys, but also make new discoveries, maybe even rethink their whole project. They expect much of their early writing to end up in their files or wastebasket, and so they start early enough to leave time for revision. That's when they figure out whether what they think they know is what they finally can say.

In chapter 12 we walk you through the process of planning and drafting a paper, then, in chapter 13, through the equally demanding task of revising its organization. In chapter 14 we explain how to frame your whole project with an introduction that motivates your readers to read carefully. In chapter 15 we discuss how to present complex data in visual form clearly. Finally, in chapter 16 we deal with problems of writing in a clear and direct style.

Deadlines always come too soon, leaving us longing for another month, or week, or even just one more day. (The three of us fought deadlines for this book, when we first wrote it and again with this edition.) In fact, some researchers seem never to be able to finish; they think they have to keep working until their report, article, dissertation, or book is perfect. No such perfect document exists, ever has, or ever will. All you can do is to make your report as complete and as close to right as you can, given the time available. When you do that, think to yourself:

> After my best efforts, here is what I believe to be true—not the whole or final truth, but a truth important to me and I hope to you, a truth that I have supported as fully as time and my abilities have allowed, so that you might find in my argument good reason to consider it, even to accept it, and perhaps even to reconsider what you believe.

An outline can be useful, but also a nuisance. The three of us remember as students grumbling over writing those outlines with roman numerals, each level indented, no subhead "A" without a "B." (Of course, we usually wrote the paper first, outlined it, then claimed we did it the other way around.) Some writers need a detailed outline; for others, a formal outline is too detailed. Booth has a file of twenty-one outlines that over seven years guided the writing of one of his books. Colomb used almost as many for his first book, but with every new one, he made it a point to discard the old. Williams outlined his book on the history of the English language on the back of an envelope. In fact, different kinds of outlines are useful at different stages in the process: the key is to distinguish *topic-based* and *point-based* outlines and know when each is useful.

A topic-based outline consists of nouns or noun phrases, not statements.

I.   Introduction: Computers in the Classroom
II. Uses of Computers
     A. Labs
     B. Classroom Instruction
III. Revision Studies
     A. Study A
     B. Study B
IV. Conclusions

A sketch like that can guide your earliest drafting, if you think you know the point you want each part to make.

But once you start drafting, you need to see your paper not as a series of topics but as a series of points (your main reasons). To do that, you need a point-based outline:

I.   Introduction: Value of classroom computers uncertain.
II. Different uses have different effects.
     A. All uses increase flexibility.
     B. Networked computer labs allow student interaction.
     C. Classroom instruction does not enhance learning.
III. Studies show that the benefits on revision are limited.
     A. Study A: writers more wordy.
     B. Study B: writers need hard copy to revise effectively.
IV. Conclusion: Too soon to tell how much word processors improve learning.
     A. Too few reliable empirical studies.
     B. Too little history, too many programs in transition.

That kind of outline helps you see whether your points hang together. You might not be able to create one until you have a draft, but the sooner you can make one to test your argument, the better. If you have a visual bent, lay out your outline as a storyboard, with each point and its support on a separate card or page (see p. 139). Many writers grasp structures better when they can literally see them laid out on a table or wall. (Storyboards are especially useful when you write in groups.)

CHAPTER TWELVE

Planning and Drafting

If you have followed our advice, you have already done a lot of writing that should help you begin a first draft. But if you are having trouble getting started, this chapter should help, whether you are on your first or twentieth research project.

Nothing is easier than putting off a first draft: *Just another week of reading,* you think, *another day, an hour; as soon as I finish this cup of coffee.* And in the long run, nothing is more likely to cause you grief. Writing is hard work, harder than reading one more article, or even ten. But you have to start sometime, and you'll start more easily if you plan carefully. In fact, you've already started if you've assembled the elements of your argument as we suggested in the last several chapters. From that plan, it's a short step to a first draft.

12.1 PRELIMINARIES TO DRAFTING

If you think before you write and then sketch a tentative plan, drafting goes faster and produces a better result. Those who just sit down and try to think of the first word and then the next risk writer's block or, worse, a trickle of words that puddle into shapeless paragraphs. But before you plan a draft, you should first reflect on both *why* and *how* you draft.

12.1.1 Exploratory Drafting vs. Planned Drafting

Experienced writers often begin to write before they know exactly what they intend to say, but they also know that their first draft is only exploratory and that much of what they write will not survive. So they start early and plan on lots of revising. The risk in unplanned exploratory writing is that you may not move on to serious drafting before a deadline forces you to hand over whatever you have. Exploratory drafting can help you discover things you never imagined, but it is not efficient if you have time for only a draft or two. If you must get a final draft out quickly, you have to draft purposefully.

12.1.2 Two Styles of Purposeful Drafting

QUICK AND DIRTY. Once they have a plan, many writers draft as fast as they can make pen or keys move. Not worrying about style or even clarity, and least of all perfect grammar and spelling, they try to keep up the flow of ideas. If a section bogs down, they note where, check their outline, and move on. If they are on a roll, they don't bother typing out quotes or footnotes: they cite just enough to know what to add later. Then if they freeze up, they have things to do: add quotes, fill in long quotations, make sure the bibliography includes every source—whatever diverts them from what is blocking them but keeps them on task, giving their subconscious a chance to work on the problem. Or they take a walk.

SLOW AND CLEAN. Others can write only sentence by polished sentence. If you cannot imagine a quicker but rougher style of drafting, don't fight it. But remember: The more small pieces you nail down early, the less you can move them around later. If you try to make large-scale revisions, you'll face a big problem, because even a minor change may require more collateral changes than you have time for. So if you draft sentence by perfect sentence, create a detailed outline that tells you where you are going and how to get there, then regularly monitor your progress.

Whatever your style, create a ritual for writing. Set daily time commitments and page goals. Ritualistically straighten up your desk, sharpen your pencils or boot up your computer, get the light just right. Don't check e-mail or start up your browser. Resolve that you will sit there writing for at least a minimum time, whether the words that come seem brilliant or dull.

12.2 PLANNING: FOUR TRAPS TO AVOID

Beginning researchers often have problems organizing a first draft because they are learning how to write as they are discovering what to write. In that struggle, they grasp the first principle of organization that comes to mind. We will describe some reliable plans in a moment, but first we explain four plans to avoid, because they shut down original thinking.

12.2.1 Don't Organize Your Report Around Your Assignment

Beginning researchers often map their papers onto the literal wording of their assignment. Do that *only* if your assignment requires it and *only* if you can think of no better way. If the assignment asks you to compare A and B, don't assume that your report must have two parts, one for A and one for B, in that order. If you echo the assignment word for word in your first paragraph, your teacher is likely to think that you have contributed no ideas of your own, as in this example.

Instructor's Assignment:

Different theories of perception give different weight to cognitive mediation in processing sensory input. Some claim that input reaches the brain unmediated; others that receptive organs are subject to cognitive influence. Compare two theories of visual, aural, or tactile perception that take different positions on this matter.

Student's Opening Paragraph:

Different theorists of visual perception give different weight to the role of cognitive mediation in processing sensory input. In this paper I will compare two theories of visual perception, one of which . . .

12.2.2 Don't Just Summarize Sources

When you are unfamiliar with a subject, it's easy to string together summaries and quotations, especially if you begin with "background." The worst form of this is called "quilting," stitching together quotations from dozens of sources in a design that reflects little of your own thinking. It invites the complaint *This is all summary, no analysis.* Some fields require you to survey what others have written, but your instructor will look for *your* angle in those summaries, for evidence of *your* mind working on those sources.

Quilting is a particular risk if you do most of your research on the Internet. After you download a few quotations, graphs, tables, and charts, nothing is easier than pasting them together with a few transitions. But experienced teachers now recognize a collage of Web screens, so you won't get anything like that past most of them.

If you are doing field research, do not simply report observations or repeat quotes from interviews. Here, too, your own contribution should appear *throughout* your report in the way you select and use your data. Use observations to support your analysis, not as a substitute for it.

12.2.3 Don't Structure Your Report Around the Topics of Your Data

It is tempting to organize a report around obvious topics such as the people or things you write about. But it is better to arrange data into more analytical categories created by your own study of the topic. Suppose you are writing about dreams and the imagination in Freud and Jung, focusing on social and biological variables. You might jump to the obvious organization—first-half Freud, second-half Jung—because their names are recognizable and the data can be divided into two simple categories. But that order prevents readers from seeing how you have analyzed your data in the context of your claim. If you claim *Dreams depend more on biological variables, the imagination more on social variables,* then organize your report not around Freud and Jung or even around social and biological variables, but around dreams and the imagination.

12.2.4 Don't Structure Your Report Around a Story of Your Research

Few readers are interested in a blow-by-blow account of what you found, then the obstacles you overcame, then a new lead you pursued, then how you hit on an answer. You see signs of this problem in language like *The first issue that I addressed was . . . Then I compared . . .* Highlight every sentence that refers to the conduct of your research rather than to its results, or to your acts of thinking and writing rather than directly to your ideas. If you highlight more than a few such sentences, you are probably not supporting a claim but telling a story about how you found it. Cut what does not help readers grasp your argument, then reorganize what remains. In laboratory research, don't bury your results in a step-by-step narrative of your lab work. When you do describe your method, your contribution must appear in how you selected only relevant details.

12.3 A PLAN FOR DRAFTING

In general, before you settle on an outline, spend time categorizing and recategorizing your data to help you find a point of view that best reflects and helps your readers see your thinking. What categories best reflect the structure of your reasoning? Use them to organize your argument. You might even discover a claim more interesting than the one you at first proposed. As you read about these next steps, do not assume you have to take them in a fixed order; go through them all, but in a way that suits your own needs.

12.3.1 Decide Where to State Your Main Point

If you have a sense of your main claim, express it as specifically as you can, then decide where to state it for the first time. Practically speaking, you have only two choices:

- in your introduction, at or close to its end;

- in your conclusion, at or close to its beginning.

This choice is crucial, because it creates your social contract with your readers. If you state your main point (what we called your main *claim* when we were discussing the elements of your argument) at the end of your introduction, you say to readers: *You now control your reading of this report. You know the outlines of my problem and its solution. You can decide how—or even whether—to go on.* Readers will feel no suspense about your claim, but if it is a serious one, they will want to know how you support it. On the other hand, if you wait until your conclusion to state your main point, you establish a relationship that is more controlling: *Readers, you will follow me through this report, considering every item I offer in an order of my choosing, until the end, where I will reveal my conclusion.* You force readers to figure out where your evidence is leading them, as they would in a murder mystery.

Most readers of research reports (or of most nonfiction writing, in fact) prefer to see the main point early, at the end of the introduction, because that puts them in control of their reading and helps them understand better the relevance of everything that follows. In many fields, in fact, the standard forms require a point-first structure, including an abstract that presents the main point and summarizes its support. (Some abstracts, however, summarize only the problem and methodology; see pp. 219–21.)

In other fields, however, the standard forms require you to put your main point in a section called *Conclusion.* If you follow that form, readers still need to know from the beginning where your report is heading. So in your introduction give them a sense of direction (of course, they are likely to flip to your conclusion, read that, and start over, or put your paper aside).

POINTS, CLAIMS, SOLUTIONS, ANSWERS, AND OTHER TERMS FOR YOUR MOST IMPORTANT IDEAS

In part II we used the terms *answer* and *solution* to refer to the sentence or sentences that resolved your central issue. In part III we used the term *claim* to refer to the sentence or sentences that make the central assertion that your argument supports. The answer, solution, and claim are usually the same sentence. Those terms also refer to the main *point* of your report (some also use the term *thesis*). We use different terms for the same sentence because each term defines it from a different angle. Most papers, whether they've involved research or not, make *points*—first, a main point central to the whole and then subpoints central to each section and paragraph. In a research report, your main point is also your *main claim* and your subpoints are also *reasons.* Your main point/claim also answers your question or states the solution to your problem.

12.3.2 Plan a Working Introduction

Some writers wait until they've written the last words of a report before they write the first ones, but, along with a plan, most of us need a working introduction to start us in the right direction. You should expect to revise it, maybe even discard it, but for your immediate purposes, it should be as explicit as you can make it (glance back at pp. 129–31).

The least useful working introduction announces only a thin topic:

> This study is about birth order and success among recent immigrants.

Better to create a *brief* context, then succinctly state your question and why it's important, followed by its answer, if you know it. If you don't have a solution, at least try to characterize the kind of solution you hope to find. Here are those four steps distinguished:

> First-born middle-class native Caucasian males are said to earn more, stay employed longer, and report more job satisfaction.*context*
>
> But no studies have looked at recent immigrants from Southeast Asia to find out whether they repeat that pattern. If it doesn't hold, we have to understand whether another does, why it is different, and what its effects are,*question*
>
> because only then can we understand how patterns of success and failure in ethnic communities differ from those in European communities.*consequences of question*
>
> The predicted influence seems to cut across ethnic groups, particularly those from Vietnam, though it partly depends on how long a family has been here and their economic level before they came.*a hypothetical answer*

That introduction barely sketches the problem and nods toward a solution, but it is enough to start you on track. In your last draft, you will revise it to state a clearer and more complete idea of the problem and solution you propose (we'll look at introductions in detail in chapter 14).

If you are really stuck for a way to begin, go back to chapter 4 (pp. 64–66) and paraphrase that three-step framework:

> This study examines the correlation between economic success and birth order among recent male immigrants from Southeast Asia $_{topic}$ to determine whether the same pattern that holds in native-born males holds with them. $_{question\ 1}$ This will allow us to explain how family background and ethnicity influence social mobility across cultures. $_{question\ 2}$

If you decide to announce your main point early, state it here, at the end of your introduction. Then check back as you draft to make sure your argument still supports it. If you decide to hold off your point until your conclusion (you had better have a good reason for doing so), end this working introduction with your point anyway. Then when you revise, delete it and substitute a "launching point," something we'll discuss in chapter 14.

12.3.3 Organize the Body of Your Report

In some fields you can map your argument onto a conventional, prescribed organization. A typical order for an experimental paper is *Introduction—Methods and Materials—Results—Discussion—Conclusion*. If readers in your field expect such forms, you have few choices about how to organize your argument. In other fields, however, you have to find your own organization. Here is a standard plan suitable for a report in which you have put your main claim at the end of your introduction:

1. Sketch necessary background, definitions, and conditions. Once you have a working introduction, decide what your readers must know and understand *before* they can understand the substance of your argument. Depending on your field, you may have to spell out your problem in more detail, define terms, review prior research, establish important warrants, set limits on your project, locate your problem in a larger historical or social context, and so on.

Keep this background short, or you risk aimless and irrelevant summary. *Do not let this summary dominate your report.* Do not summarize in detail the plot of a play or novel, all the recent research on your topic, the historical background of an event, and so on. Present only enough background for readers to understand special terms, research that motivated yours, and basic facts about your topic. If your background section is more than a few pages long (and that may already be too long), end it with a concise statement of what you want your readers to carry with them as they begin the main body of your argument. Then consider replacing that full section with its summary.

2. Find the best order for your reasons and evidence. Some fields require a report based on experiments to have a separate section called *Results* or *Findings*. That's where you report your data. If, however, you have to plan a report around a sequence of reasons, you normally state your reason first, then lay out the evidence supporting it. Some writers lay out their evidence first, as a kind of mystery story, but the default order is reason first, then evidence.

Far trickier is finding the best order for your reasons. If your argument depends on a sequence of parallel reasons, try arranging them in different orders. It costs less to reject bad choices now than to revise them later. You can do this most easily if you print out your outline as a storyboard, with each main reason at the top of a separate page. Add supporting material, particularly evidence, letting it run onto more pages if necessary (staple them together). Then try out different combinations and orders, keeping in mind the needs of your readers. You can rely on a few principles that turn on what your readers know and understand.

- **Old to New.** In general, readers prefer to move from what they know to what they don't. Take this principle as a general guide when you are stuck: Start with what's familiar *to your readers,* then move to the unfamiliar.

- **Shorter and Simpler to Longer and More Complex.** In general, readers also prefer to deal with shorter, less complex reasons before longer, more complex ones. Start with the elements of your argument that readers will understand most easily. The easiest parts are likely to be more familiar as well.

- **Uncontested to More Contested.** In general, readers move more easily from less contested to more contested issues. If your main claim is controversial and you can present several arguments to support it, try starting with the one your reader is most likely to accept.

Consider these possible orders, as well:

- chronological order;

- logical order, from evidence to reason to claim, or vice versa;

- concessions and conditions first, then an objection you can rebut, then your own affirmative evidence, or vice versa.

Unfortunately, these criteria can pull against one another: what some readers understand best are the objections they hold most strongly; what you think is your most decisive argument can be the newest and most contested claim. We can offer no absolute rules here, only variables to consider. Presiding over all your judgments must be this principle: What must your readers know before they can understand what comes next?

Incidentally, the principle about stating points early also applies to major sections and subsections. Readers prefer to find the main point of a section in its first few sentences. If they don't, they still need an introductory sentence or two to frame what they are about to read.

FINDING THE RIGHT ORDER

Expect to try out several orders before you find the right one. We did. The chapters you are reading are ordered differently from the first edition, because some readers told us that the original order didn't "flow." Among other changes, we moved the chapter on the most difficult topic, warrants, to the end of part III, so that if readers got discouraged in that chapter, they would at least have finished reading about the other parts of argument first. But changing that order was nothing new: we had already tried out more than a dozen orders in drafting the first edition (and still didn't get it quite right).

3. Locate acknowledgments and responses. Try to acknowledge and respond to the most important questions and objections where you think readers will raise them.

- If your *whole* argument counters someone else's, *briefly* summarize that other argument in your introduction. Then, in the body of your argument, develop it in more detail. Once you lay it out, you can work through it, step-by-step.

- If you think an alternative will occur to your readers as they read, but you want to finish a point before you acknowledge it, *briefly* acknowledge the alternative, finish your point, then return to respond.

- If you think an alternative will occur to readers only after they understand something, offer it there.

- If there are several alternative possible solutions to your problem, you can plan your argument by sequentially posing and eliminating them, leaving your solution as the last one standing.

  How do we deal with global warming? Some suggest we ignore it. . . .*explanation* But that won't do because . . .

  Others say we exploit it by adapting our lives to warmer conditions. . . .*explanation* But that won't work either because . . .

  At the other extreme, some argue we should end all CO_2 emissions. . . .*explanation* But that is impractical because . . .

  None of these addresses the problem responsibly. The only solution is to . . .

4. Locate warrants. Once you have determined whether to state warrants, you have to decide where to put the ones you will. Generally speaking, state a warrant before you offer your claim and its supporting reason. What's missing in this example?

> Since most students at Oxford University in 1580 added nothing to their signatures,$_{reason}$ most of them must have been commoners.$_{claim}$

Unless you are an expert in Elizabethan social history, that little argument makes no sense. It is clearer to everyone (even experts) when it is introduced by a warrant:

> **In late-sixteenth-century England, only those few men called "gentlemen" could sign their names with an added "Mr.," and only the son of a gentleman could add an "Esq."**$_{warrant}$ Since most students at Oxford University in 1580 added nothing to their signatures,$_{reason}$ most of them must have been commoners.$_{claim}$

If you think you need a warrant to make your main claim clearly relevant to your most important reason, state it early in your report, and if you think readers might question it, make an argument supporting it.

You can also state a warrant after a specific claim and supporting reason, as a kind of rhetorical flourish that seems to wrap up the argument on an emphatic note:

> We should have suspected all along that Thomas Jefferson had a relationship with his slave Sally Hemings,$_{claim}$ if only because there were so many contemporary reports of one.$_{reason}$ **After all, where there's smoke, there's likely to be fire.**$_{warrant}$

By this point, you have probably sifted out much of your data because they will seem irrelevant. That does not mean you wasted time collecting them. Research is like gold mining: dig up a lot, pick out a little, discard the rest. Even if all that material never appears in your report, it is the tacit foundation of knowledge on which your argument rests. Ernest Hemingway once said that you know you're writing well when you discard stuff you know is good—but not as good as what you keep.

12.4 THE PITFALL TO AVOID AT ALL COSTS: PLAGIARISM

It will be as you draft that you risk the worst that can happen to a researcher. You are filling up pages or your screen with lots of good words *and you forget that you collected those words from someone else.* Few researchers intentionally plagiarize, but every honest one still needs to give it serious thought, because most plagiarism is inadvertent. Sometimes it happens when a writer is not clear about what to cite or when. (If you're not sure, ask your teacher for guidance.)

Most writers who plagiarize inadvertently do so because they took notes carelessly (review pp. 91–104). The eminent historian Doris Kearns Goodwin was publicly humiliated when it was discovered that she had copied into her books hundreds, maybe thousands of words written by others. In defense, she claimed that in her note taking, she had neglected to identify the quotations as quotations. A few accepted her defense; many did not. If someone as celebrated at Goodwin can plagiarize inadvertently, every writer should strive to avoid it.

DELIBERATE PLAGIARISM IS STEALING

Students may not recognize the full damage done by deliberate plagiarism—a matter we address in part V. But it would seem impossible for them not to know that they plagiarize when they buy a paper or copy another's work and present it as their own. But in a world where we move data around so easily, students get odd ideas of ownership. Colomb once confronted two students who turned in identical papers for two sections of the same class. Faced with the evidence, one admitted that she had copied a paper shown to her by the other. Hearing this, the other student became irate, complaining that the first student had no right to copy his paper. Only it turned out that he had gotten the paper out of his fraternity's files. In his mind, what made it "his" paper was that only his fraternity brothers had the right to turn in those papers as their own!

12.4.1 Plagiarism Defined

You plagiarize when, intentionally or not, you use someone else's words or ideas but fail to credit that person, leading your readers to think that those words are yours. There are, however, complications, because different fields draw the line between fair use and plagiarism in different places. In all fields, you plagiarize when you use a source's words or ideas without citing that source. In most fields, you plagiarize even when you *do* credit the source but use its exact words without using quotation marks or block indentation. But in the law, it is acceptable to use the exact words of a court's ruling without quotation marks, if you cite it. In other fields, you plagiarize when you paraphrase a source so closely that anyone putting your work next to it would see that you could not have written what you did without the source at your elbow. But in many sciences, researchers commonly report another researcher's results using words very similar to the original. So, if you don't know the standards for plagiarism in your field, play it safe and credit the original as fully as possible.

12.4.2 Avoiding the Straightforward Plagiarism of Words

Every time you use the exact words of a source, stop. Then

- type quotation marks before and after them, or create a block quotation (see the Quick Tip at the end of this chapter);

- record the words *exactly* as they are in the source (if you change anything, use square brackets and ellipses to indicate changes);

- cite the source.

If you omit the first or last step, intentionally or not, you plagiarize.

12.4.3 Avoiding the Plagiarism of Ideas

You also plagiarize when you use someone else's ideas but do not credit that person. You would plagiarize us, for example, if you wrote about problems using our concepts from chapter 4 but did not credit us, even if you changed our words, calling conditions, say, *predicaments*, and costs, *damages*. If you base several pages on the work of another, don't just mention that fact in a footnote at the end (as one researcher did with some of Williams's and Colomb's work): attribute their work up front.

A tricky situation arises when you get an idea on your own, then discover that someone else thought of it first. In the world of research, priority counts not for everything, but for a lot. If you do not cite that prior source, you risk having people think that you plagiarized it, even though you did not.

An even trickier situation is when you use ideas that are widely known in your field, as we inevitably do here. (How could we cite the thousands of sources for our commonplace claim that your essays should be well organized?) Sometimes the idea is so familiar that everyone knows who gets credit for it, and you would be thought naive if you cited it. For example, you might mention Crick and Watson when you talk about the helical structure of DNA, but you would not cite their article announcing that discovery. At other times, you might know that an idea is common knowledge, part of the background in your field, but not know who first published it. Since you can't cite, much less track down everything you write, these are cases where even scrupulous students can misstep. All we can say is *When in doubt, ask your teacher and give credit where you can.*

12.4.4 Indirect Plagiarism of Words

It is trickier to define plagiarism when you summarize or paraphrase. They are not the same, but they blend so seamlessly that you may not be aware when you drift from summary into paraphrase, then across the line into plagiarism. No matter your intention, close paraphrase may count as plagiarism, *even when you cite the source.*

For example, this next paragraph plagiarizes what you just read, because it paraphrases it so closely:

> It is difficult to define plagiarism when summary and paraphrase are involved, because while they differ, their boundaries blur, and a writer may not know when she is summarizing, paraphrasing, or plagiarizing. Regardless, too close a paraphrase is plagiarism, even when the source is cited.

The next example is borderline:

> It is hard to distinguish summary, paraphrase, and plagiarism.
> You risk plagiarizing when you paraphrase too closely, even
> when you never meant to plagiarize and you cite a source.

The words in those versions track the original so closely that a careful reader can see that the writer could have written them only while *simultaneously* reading the original. Here is a summary of that paragraph, just on the safe side of the border:

> According to Booth, Colomb, and Williams, writers sometimes
> plagiarize unconsciously because they think they are summariz-
> ing, when in fact they are too closely paraphrasing, an act that
> counts as plagiarism, even when done unintentionally and when
> sources are cited (p. 203).

In fields that use a lot of direct quotation, such as history and English, close paraphrasing is risky.

Here is a simple way to avoid inadvertent plagiarism: Paraphrase your source, *only after those words have filtered through your own understanding of them*. Then as you actually write, take your eyes away from your source and look at the screen or your own page.

12.5 THE NEXT STEP

The biggest difference between experienced writers and beginners is their attitude toward their first draft. Beginners take it as a triumph (which it is): *Done! I'll change that word, fix this comma, run the spell checker, and <Print>!* Experienced researchers take a first draft as a challenge: *I have the sketch; now comes the hard but gratifying work of discovering what I can make of it.* They know that now, more than ever, they have to try to see their report as their readers will. And that means revising that first draft (or second or third . . .), the subject of our next chapter.

In every field, researchers have to report what other researchers write. But the practices of your particular field determine how you do that.

HOW TO QUOTE AND PARAPHRASE

In the natural sciences and some social sciences, researchers rarely quote sources directly; they paraphrase and cite them. They make the name of the source a direct part of their own sentence only if the source is important and they want to emphasize it.

> Several processes have been suggested as causes of the associative-priming effect. In their seminal study, Meyer and Schvaneveldt (1971, p. 232) suggested two: *automatic (attention-free) spreading activation* and *location-shifting*. More recently, a further associative-priming process has been studied (de Groot 1984).

The writer thought that Meyer and Schvaneveldt were important enough to name directly, but cited de Groot only in parentheses, as a minor reference.

In the humanities and some social sciences, researchers sometimes paraphrase sources, but more often quote them. You have three options:

1. Introduce a quotation with a colon or an introductory phrase.

> Plumb describes the Walpole administration in terms that remind one of the patronage system in U.S. cities: "Sir Robert was the first English politician to understand how to use the loyalty of people whose only qualification was his sponsorship" (p. 343).

> Plumb describes the Walpole administration in terms that remind one of the patronage system in U.S. cities. He claims that

"Sir Robert was the first English politician to understand how to use . . ."

2. Weave the quotation into your own sentence, making sure that the grammar of your sentence fits the grammar of the quotation.

Plumb speaks in terms that remind one of the patronage system in modern U.S. cities when he describes how Walpole was able "to use the loyalty of people whose only qualification . . ."

Jameson was never comfortable with the decisions of the Tribunal, and he often "complain[ed] . . . that something had to be changed" (1984, p. 44).

Note that square brackets indicate insertions and ellipses indicate words dropped.

3. Set off quotations of three or more lines in a "block quote." When you do, make sure the quotation connects to what has gone before, and that just before or just after the quote, you make clear why you are quoting it.

After 1660 English moralists complained that people were motivated by material goods, which was, of course, nothing new. But they noted a new form of "mercenary virtue" that offered material incentive for good behavior. Shaftesbury wrote:

> Men have not been contented to show the natural advantage of honesty and virtue. . . . They have made virtue so mercenary a thing, and have talked so much of its rewards, that one can hardly tell what there is in it, after all, which can be worth rewarding. (p. 135)

WHEN TO QUOTE AND WHEN TO PARAPHRASE

We can't give you rules for when and how much to quote or paraphrase. Quote or cite too often, and you seem to offer too little of your own work; quote too little, and readers may think your claims lack support or may not see how your work relates to that of others. There are, however, some rules of thumb.

Use direct quotations

- when you use the work of others as primary data;

- when you appeal to their authority;

- when the specific words of your source matter because
   —those words have been important to other researchers;
   —you want to focus on how your source says things;
   —the words of the source are vivid or significant;
   —you dispute your source and want to state her case fairly.

Paraphrase sources

- when you are more interested in findings and data than in how a source expresses them;

- when you can say the same thing more clearly.

Don't quote because it's easier or you think you lack the authority to speak for your sources. Make your own argument with your own claims, reasons, and evidence.

CHAPTER THIRTEEN

Revising Your Organization and Argument

The plan for revising in this chapter may at first reading seem mechanical and very detailed. But if you take each step one at a time, you can analyze your paper more easily and more thoroughly than by just reading and wondering whether it all hangs together.

Now you face a task that vexes every writer—deciding whether your report will make sense to readers. The problem is, it will always make more sense to you, because you remember too well what you meant when you wrote it. So you need ways to find out where readers are likely to struggle and then to revise it to meet their needs. As rhetorical theorists since Aristotle have said, an effective speaker or writer must "accommodate to his audience," particularly to their inability to read your mind.

Some writers resist that idea, fearing that accommodating readers compromises their integrity. They think the truth of their discovery should speak for itself. But new knowledge is never just discovered, presented, and accepted. New ideas are always created, then *shaped* to meet readers' needs, beliefs, and objections. That doesn't mean pandering to readers—if you imagine yourself having an amiable conversation with them, trying to understand what and how they think but also holding firm on what you know. You not only make your ideas clear to them; you also discover the best that you yourself can think.

To do that you need a plan for revising that keeps you in conversation with your imagined readers. To create that conversation, you cannot read your report straight through, sentence by sentence, changing a word here, fixing a spelling there. You need a more disciplined approach that tests and questions your report as your readers will, one aspect at a time. In this chapter we'll discuss how to diagnose and revise your overall organization and argument so that readers get that sense of a whole. In the next, we'll discuss how to create an introduction that frames your report and "sells" your readers on its significance.

13.1 THINKING LIKE A READER

Readers do not read sentence by sentence, accumulating information as they go, as if they were fingering beads on a string. They need to begin reading with a sense of the whole and its structure and, most important, why you want them to read the report in the first place. Since readers think the whole is more important than its parts, it makes sense to diagnose the largest elements first. Start with the overall organization, then turn to local organization, then to the clarity of sentences, and last to matters of spelling and punctuation. In reality, of course, no one revises so neatly. All of us revise as we go, correcting spelling as we rearrange our argument. But when you revise top-down deliberately, from global structure to sections to paragraphs to sentences to words, you are more likely to discover more useful revisions than if you start at the bottom with words and sentences and work up.

But since you cannot read your own writing as your readers will, you need some formal, even mechanical ways to analyze, diagnose, and revise so that you can sidestep your too-easy understanding (and too-ready admiration) of your own words. Revision is not just tidying up what you thought; it's a way of thinking anew that helps you think better, so don't rush it. In fact, this final stage is when you will understand your project most completely. Thoughtful revision takes time, so start early.

13.2 Analyzing and Revising Your Overall Organization
This process consists of three steps:

1. Identify the outer frame of your paper: your introduction and conclusion, and a sentence in each that states your main claim, your point, the solution to your problem, the answer to your question.

2. Identify the major sections of the body of your paper; then mark off their introductions and the point sentences for each of those sections.

3. In the introduction to the whole paper, identify your key thematic concepts, then track them through the rest of the paper. Then do the same for each section (this is the part that will seem very detailed).

Each of those three steps consists of further steps.

13.2.1 Step 1: Identify Your Outer Frame and Main Points

Readers must know three things unambiguously:

- where your introduction ends and the body of your paper begins;

- where the body of your paper ends and your conclusion begins;

- which is the main-point sentence in your introduction (if you put one there), and which is the main-point sentence in your conclusion.

To be sure that your readers will unambiguously identify those elements, do the following:

1. Always start a new paragraph after your introduction and another at the beginning of your conclusion. In fact, put an extra space after your introduction and before your conclusion. If your field approves, put headings at these joints to make sure your reader can't miss them.

2. In your working introduction, underline the sentence that comes closest to stating your main claim (do not count a topic-announcing sentence like *This paper will discuss . . .*).

3. In your conclusion, underline the sentence that best expresses the main point of your argument, the gist of the answer to your question.

Now compare the point in your introduction and the point in your conclusion. They should at least not contradict each other. If one is more specific and contestable, it should be the one in your conclusion. If the point sentence in your introduction is vague, merely a "topic-announcing" sentence, revise it to make it more specific.

For example, consider this introductory paragraph. What is its point?

> In the eleventh century, the Roman Catholic Church initiated several Crusades to recapture the Holy Land. In a letter to King Henry IV in the year 1074, Gregory VII urged a Crusade but failed to carry it out. In 1095 his successor, Pope Urban II, gave a speech at the Council of Clermont in which he also called for a Crusade, and in the next year, in 1096, he successfully initiated the First Crusade. In this paper I will discuss the reasons that these popes gave for a Crusade.

Here is the concluding paragraph. What is its point?

> As we can see from these documents, Popes Urban II and Gregory VII urged the Crusades as a way not just to restore the Holy Land to Christian rule, but also to preserve the political unity of the Church and western Europe. Urban wanted to conquer the Muslims, but no less importantly to reinforce his authority and control fighting among Europeans by directing their energies elsewhere. Gregory wished to unify the Roman and Greek Churches, but also to prevent the breakup of the Catholic Church and even the Empire. To achieve their political ends, each pope tried to unite people in a common religious fight against the East to prevent them from fighting among themselves and to unify an increasingly divided Church. Thus the Crusades were not just a religious effort to recapture the Holy Land and to save God's faith, as is so widely believed in popular memory, but shrewd political efforts to unify the Church and Europe and save them from internal forces threatening to tear them apart.

The point sentence in the introduction appears to be the last, rather vague one:

> I will discuss the reasons that these popes gave for a Crusade.

But that sentence merely announces, *I'm going to tell you about the Crusades.* The point sentence in the conclusion seems to be the last sentence:

> Thus the Crusades were . . . shrewd political efforts to unify the Church and Europe and save them from internal forces threatening to tear them apart.

That point is more specific, more substantive, and plausibly contestable. Now we know how to revise the last sentence of the introduction. We could just cut and paste that concluding sentence into the introduction, or we could write a sentence that, while not revealing the full point, would at least connect the two more clearly, like this:

> In a series of documents the popes proposed their Crusades to restore Jerusalem to Christendom, but their words suggest other issues involving political concerns about European and Christian unity in the face of forces that were dividing them.

13.2.2 Step 2: Identify Your Major Sections and Their Points

There are three more things that readers must unambiguously know about your organization:

- where each section of your report stops and the next begins;

- how each subsequent section relates to the previous one;

- what the main point of each section is.

So in each section of your report, do what you just did for the whole paper.

1. Divide the body of your paper into its major sections. Put an extra space between them. If you cannot find section boundaries, your readers won't either.

2. Put a slash mark after the introduction to each major section. Each section must have a short segment that introduces it.

3. Put a slash mark before the conclusion to each major section. (If a section is shorter than a page or so, it may not need a separate conclusion.)

4. In each section, highlight the sentence that expresses its point (probably one of your argument's main reasons). If you cannot find it, neither will your readers. Ordinarily, it should be at the end of a brief introduction to that section. If it is not, you must have good reason for putting it at the end. (*Never* locate the only point of a section in its middle.)

5. Underline the first several words of every new section. They should signal how that section relates to the previous one. Readers have to understand why sections are in the order they are. If they don't, they judge what they read to be incoherent.

If you can't do each of those steps quickly, you have uncovered a problem with the organization of your report. Look again at pp. 138–40 and 191–200 to review how you arranged your ideas and structured your argument.

When you highlighted your points, you made an outline that you can now read off the page, though you would do better to write it out. It will be a list of sentences that might look like the one on p. 188. Now ask yourself, *If I tried to assemble these sentences into a single paragraph, would it make sense?* If it doesn't, you face a big problem.

13.2.3 Step 3: Diagnose the Continuity of Your Themes

Your next step is to determine whether these points and subpoints "hang together" not just linearly but conceptually. This is detailed work, but worth the time and effort. You must track the words that express your key concepts from your introduction through the body of your report into your conclusion. If readers do not see those key themes throughout your paper, they may think your report is unfocused.

Do this:

1. In your introduction and conclusion, particularly in their point sentences, circle key concepts (we'll call these *themes*). Ignore general words like *topic, issue, important, significant,* and other words that do not refer to the content of your claim.

2. If you cannot find keywords in your introduction or find only a few,

   - look closely at the last few pages of your report for the concepts that appear there most often;

   - incorporate those concepts into your two point sentences, both at the end of your introduction and in your conclusion.

For example, in the Crusades example above, the point sentence in the introduction is empty of significant concepts:

> I will discuss the reasons that these popes gave for a Crusade.

There are, however, several key terms in the last paragraph (and several more in the preceding ones):

> As we can see from these documents, Popes Urban II and Gregory VII urged the Crusades as a way not just to restore the Holy Land to Christian rule, but also to **preserve** the **political unity** of the Church and western Europe. Urban wanted to conquer the Muslims, but no less importantly to **reinforce his authority** and to **control fighting** among Europeans by directing their energies elsewhere. Gregory wished to **unify** the Roman and Greek Churches, but also to **prevent** the **breakup** of the Catholic Church and even the Empire. To achieve their **political** ends, each pope tried to **unite** people in a **common** religious fight against the East to **prevent** them from **fighting among themselves** and to **unify** an increasingly **divided** Church. Thus the Crusades were not just a religious effort to recapture the Holy Land

and to save God's faith, as is so widely believed in popular memory, but shrewd **political efforts** to **unify** the Church and Europe and **save** them from **internal forces** threatening to **tear them apart.**

We can assemble the key conceptual themes under just a few terms:

preserve political/religious unity, prevent fighting and breakup, political efforts

These are the terms that should be built into a new point sentence for the introduction.

If the circled terms in your conclusion are more detailed than those in your introduction, your introduction may be too vague to give your readers a sense of where you are taking them. They must recognize the central thematic concepts that hold your paper together, and after they finish, they must have those concepts etched in their memory. If they don't, they may feel that you have gotten off track or have broken the implied promise in your introduction.

The next step is to trace whether those circled key terms appear consistently in the subpoint sentences in the rest of your outline. We do not have the space to illustrate that step, but you can do for each section what we did with the introduction and conclusion to the paper about the Crusades:

1. Circle words in the subpoint sentences that are the same or related to the circled terms in the point sentences in your introduction and conclusion.

2. If any subpoint sentence has no terms from your main point, you may have failed to relate that subpoint to your main claim. Revise the subpoints to include a few circled terms. If you can't, consider revising or even eliminating that section.

3. Now do the opposite. Check for important concepts in your subpoints that you did *not* mention in the introductory or

concluding point sentences. Should you add these key terms to those point sentences?

If your field encourages headings, create them for each major section:

1. In the point sentence of each section, identify key terms that appear only or most often in that section.

2. Assemble those key terms into a heading that uniquely identifies the section.

If you find it difficult to create headings, you may have a serious problem with the coherence of your paper, and if you do, so will your readers. This is such a good diagnostic test that even if headings are not used in your kind of writing, you should try to create them, then delete them before you print out your last draft. If your report is long and you have the time, repeat this process for each major sub-subsection.

13.2.4 Diagnose the Whole Again

Once you've done all that, read all your point sentences again as if they were a single paragraph summarizing the whole report. We can't give you a surefire way to know whether they add up to a whole, so this is a good time to ask a friend, relative, or roommate to listen to an oral presentation of that summary of your report. Use your outline of points as a guide.

13.3 REVISING YOUR ARGUMENT

Once you determine that your organization is at least plausible, question whether that organization expresses an argument or only a patchwork of quotations and data.

13.3.1 Identify Your Argument

Return to your outline of main and subpoints. First check whether the organization of the paper reflects the organization of its argument:

1. Determine whether those reasons are the major claims that each section supports. If not, you have a disjunction between the organizing points of your paper and the structure of claims in your argument.

Then test how much of your discussion reflects the analysis in your argument:

2. In each section, identify everything that counts as evidence—all the summaries, paraphrases, quotations, facts, figures, graphs, tables—anything you report from a primary or secondary source.

3. Now, ignoring all of that, skim what's left. You are looking for the *expression* of your analysis, your evaluation, your judgment.

If what is left is much less than what you ignored, you may not have a substantial argument, but only a pastiche of data and summary. If you have time, return to chapters 7–11 and beef up your own contribution to your report.

13.3.2 Evaluate the Quality of Your Argument

Now ask some harder questions. Assuming that your readers can follow the organization of your argument, what might cause them to reject it? At this point, you have to evaluate your evidence, your reservations, and what is most difficult, your warrants. (If you need a review, see chapters 7–11.)

1. Is your evidence reliable and clearly connected to your claims? If you are close to a final draft, it may be too late to find evidence that is more representative or precise, and if you are using all the evidence you have, you can do nothing about its sufficiency and appropriateness. But you can check other features:

• Check your data and quotations against your notes.

• Make sure your readers see how quotations and data relate to your claim.

- Be sure you haven't skipped intermediate steps in an argument (review especially pp. 138–40).

2. Have you appropriately qualified your argument? Don't hesitate to drop into appropriate places a few well-placed hedges like *probably, most, often, may,* and so on.

3. Does your report seem less like a contest between competitors and more like a conversation with colleagues who are amiable but have minds of their own? Readers want to hear reasons, not to challenge you but because they want to know more. *Why do you believe that? But what about . . . ? Are you really making that strong a claim? Could you explain how this evidence relates to your claim?* Go through your argument, asking such questions in unexpected places (review pp. 152–57).

4. The hardest question: What warrants have you left unexpressed? Even if your readers accept that your reasons are based on reliable evidence, what else must they believe before they accept your claims (review pp. 168–70)? There is no easy way to test this. Once you have identified each section and subsection of your argument, write in the margin the most important unstated warrant that you think readers must accept. Then ask whether they will, or whether you have to argue for it explicitly.

13.4 THE LAST STEP

In the Quick Tip on speedy reading after chapter 6, we described how to skim your sources for their gist to see whether they offered anything useful. Give someone else your paper to skim in the same way, asking them to report its gist to you. If that reader can skim your paper easily and report its gist accurately, you probably have a well-organized paper. If not . . .

TITLES

The first thing your readers read—and the last thing you should revise—is your title. Most writers attach a few words to suggest what a report is "about," but a title is useful only if it helps readers understand specifically what is to come. Compare these two:

Economic Sources of the "Separate but Equal" Doctrine

Equal Rights, Unequal Education:
Economic Racism as a Source of the "Separate but Equal" Doctrine

Begin with your working title (p. 44), then refine it to make it useful to your readers. It should introduce the key themes that are in your main claim, the ones you circled when you checked for the continuity of key conceptual themes. When readers see concepts in your title turn up in the body of your paper, especially in your main claim, they will feel that the text has met their expectations. (Incidentally, two-line titles will give you more scope to specify the key terms in your report. End the first line with a colon that introduces a more specific second line.)

ABSTRACTS

In some fields, particularly in the natural and social sciences, reports begin with an abstract, a brief summary that tells readers what to expect. It should be shorter than an introduction, but still do three things that an introduction does:

- It states the research problem.

- It announces key themes.

- It ends with a statement of the main point or with a launching point that anticipates the main point in the full text.

Abstracts differ in different fields. But most follow one of three patterns. To determine which patterns are used in your field, ask your teacher or look in a standard journal.

1. Context + Problem + Main Point

This kind of abstract is an abbreviated introduction. It begins with a sentence or two to establish the context of previous research, continues with a sentence or two to state the problem, and then concludes with the main result of the research:

> Computer folklore has long held that character-based user interfaces promote more serious work than do graphical user interfaces (GUI), a belief that seemed to be confirmed by Hailo (1990).*context* But Hailo's study was biased by the same folklore that it purported to confirm.*problem* In this study, no significant differences were found in the performance of students working with a character-based interface (MS-DOS) or with a graphical interface (Macintosh OS).*main point*

2. Context + Problem + Launching Point

This pattern is the same as the previous one, except that the abstract states not the specific results, only their general nature:

> Computer folklore has long held that character-based user interfaces promote more serious work than do graphical user interfaces (GUI), a belief that seemed to be confirmed by Hailo (1990).*context* But Hailo's study was biased by the same folklore that it purported to confirm.*problem* This study tested the performance of thirty-eight business communication students using either a character-based or a graphical interface.*launching point*

3. Summary

A summary states the context and the problem, but before reporting the result, it summarizes the rest of the report, focusing either on the evidence supporting the result or on the procedures and methods used to achieve it:

> Computer folklore has long held that character-based user interfaces promote more serious work than do graphical user inter-

faces (GUI), a belief that seemed to be confirmed by Hailo (1990).*context* But Hailo's study was biased by the same folklore that it purported to confirm.*problem* In this study, thirty-eight students in the same technical communication class were randomly assigned to one of two computer labs, one with character-based interfaces (MS-DOS), the other with graphical interfaces (Macintosh OS). Documents produced were evaluated on three criteria: content, format, and mechanics.*summary* There was no significant difference between the two groups on any of the three criteria.*main point*

In years to come, some researcher may search for exactly the research you have done. That search will be done by a computer looking for keywords in titles and abstracts. So when you write yours, imagine looking for your own research. What words should a researcher look for? Put them in your title and abstract.

CHAPTER FOURTEEN

Introductions and Conclusions

This chapter builds on ideas introduced in chapter 4. We show you how to introduce your research report with a problem that motivates readers to read it and conclude it in a way that emphasizes its significance. Nothing is more useful than a strong introduction and conclusion that help readers see the significance of your work.

Once you have a revised first (or second or third) draft, you're ready to revise your working introduction so that readers know where you will take them and why they should go there. Some writers think that means following the old advice: *Grab their attention with something snappy or cute.* That advice is not useless, but those who read research reports look for more than cute. What grabs readers is a problem they think is in need of a solution, and what holds them is the hope that you've found it.

In this chapter we show you how to write an introduction that frames your report so that readers can read it faster and understand it better, because they know both what to expect and why they should care. We then show you how to conclude your report so that readers come away not only with a clear understanding of your claims but also with renewed appreciation of their significance. The time you spend on your introduction and conclusion may be the most important revising you do. As we've said, you can always work with readers inclined to say, *I don't agree.* What you can't survive are readers who shrug and say, *I don't care.*

14.1 THE THREE ELEMENTS OF AN INTRODUCTION

As we've steadily emphasized, different research communities do things in different ways, but nowhere do those differences seem more striking than in their introductions. These three, for example, are from the fields of cultural criticism, computer design, and legal history. They look different on the surface, but in fact, they are much alike in their structure:

(1) Why can't a machine be more like a man? In almost every episode of *Star Trek: The Next Generation*, the android Data wonders what makes a person a person. In the original *Star Trek*, similar questions were raised by the half-Vulcan Mr. Spock, whose status as a person was called into question by his machinelike logic and lack of emotion. In fact, Data and Spock are only the most recent "quasi-persons" who have explored the nature of humanity. The same question has been raised by and about creatures ranging from Frankenstein to Terminator II. But the real question is why characters who struggle to be persons are always white and male. As cultural interpreters, do they tacitly reinforce destructive stereotypes of what it is about a person that we must think of as "normal"? The model person, to which we all must aspire, seems in fact to be defined by Western criteria that exclude most of the people in the world.

(2) As part of its program of Continuous Quality Improvement ("CQI"), Motodyne Computers plans to redesign the user interface for its Unidyne™ online help system. The specifications for the interface call for self-explanatory icons that will allow users to identify their function without an identifying label. Motodyne has three years' experience with its current icon set, but it has no data showing which icons are self-explanatory. Lacking such data, we cannot determine which icons to retain and which to redesign. This report provides data for eleven icons, showing that five of them are not self-explanatory.

(3) In today's society, would Major John André, a British spy captured behind American lines in civilian clothes in 1780, be hanged? Though considered a noble patriot, he suffered the punishment mandated by military law. Over time, our traditions have changed, but the punishment for spying has not. It is the

only offense for which death is mandated. Recently, though, the Supreme Court has rejected mandatory death sentences in civilian cases, creating an ambiguity in their application to military cases. If Court decisions apply to the military, then Congress may have to revise the Universal Code of Military Justice. This article concludes that to be the case.

The topics and problems posed in those three introductions differ as much as their intended readers, but behind them is a shared rhetorical pattern that readers look for in all introductions, regardless of field. That common structure consists of three elements:

- contextualizing background,

- a statement of the problem,

- a response to the problem.

Not every introduction has all three of those elements, but most do, and the vast majority state at least part of a problem.

We can see that structure of *Context* + *Problem* + *Response* in all three of those introductions:

(1) OPENING CONTEXT: Why can't a machine be more like a man? . . . The same question has been raised by and about creatures ranging from Frankenstein to Terminator II.

PROBLEM: But the real question is . . . do they tacitly reinforce destructive stereotypes of what it is about a person that we must think of as "normal"?

RESPONSE: The model person, to which we all must aspire, seems in fact to be defined by Western criteria that exclude most of the people in the world.

(2) OPENING CONTEXT: As part of its program of Continuous Quality Improvement ("CQI"), Motodyne Computers plans to redesign the user interface. . . . Motodyne has three years' experience with its current icon set,

PROBLEM: but it has no data showing which icons are self-explanatory. Lacking such data, we cannot determine which icons to retain and which to redesign.

RESPONSE: This report provides data for eleven icons, showing that five of them are not self-explanatory.

(3) OPENING CONTEXT: In today's society, would Major John André . . . be hanged [for spying]? . . . It is the only offense for which death is mandated.

PROBLEM: Recently, though, the Supreme Court has rejected mandatory death sentences in civilian cases, creating an ambiguity in their application to military cases. . . . Congress may have to revise the Universal Code of Military Justice.

RESPONSE: This article concludes that to be the case.

Each of those elements of an introduction plays it own role not only in helping readers understand, but in motivating them to read. We will discuss them in their order.

14.2 ESTABLISHING COMMON GROUND

We call contextualizing information *common ground,* because it establishes a shared understanding between reader and writer about the general issue the writer will address. But it does something even more crucial, illustrated best with the opening of a fairy tale:

One sunny morning, Little Red Riding Hood was skipping happily through the forest on her way to Grandmother's house, when suddenly Hungry Wolf jumped out from behind a tree, frightening her very much.

Like the opening to most fairy tales, this one establishes a stable, unproblematical, even happy context:

STABLE CONTEXT: One sunny morning, Little Red Riding Hood was skipping happily through the forest on her way to Grandmother's house _stable context_ [imagine butterflies dancing around her head to flutes and violins].

That stable context is then disrupted with a problem:

DISRUPTING PROBLEM: . . . When suddenly Hungry Wolf jumped out from behind a tree _condition_ [imagine trombones, tubas, and bass fiddles], frightening her [and, if they lose themselves in the story, little children as well]._cost_

The rest of the story develops that problem and then resolves it.

Unlikely though it may seem, introductions to most research reports follow the same strategy. They open with the stable context of a common ground—some apparently unproblematic account of research, a statement of the community's consensus on a familiar topic. The writer then disrupts it with a problem: _Reader, you think you know something, **but** your knowledge is flawed or incomplete._

(3) OPENING CONTEXT: In today's society, would Major John André, a British spy . . . be hanged? . . . [Spying] is the only offense for which death is mandated.

PROBLEM: Recently, **though**, the Supreme Court has rejected mandatory death sentences. . . .

Not every research report opens with common ground. Here is an introduction that opens directly with a problem:

Recently the chemical processes that have been thinning the ozone layer have been found to be less well understood than once thought. We may have labeled hydrofluorocarbons as the chief cause incorrectly.

Some readers might find that problem already disturbing enough to motivate their reading, but you can heighten its rhetorical punch by introducing it with an unproblematical context of prior research, not just to orient readers toward the topic, but specifically to create an apparently stable context just so that you can disrupt it:

As we have investigated environmental threats, our understanding of many chemical processes such as acid rain and the buildup of carbon dioxide has improved, allowing us to understand better their eventual effects on the biosphere.*common ground* (*Sounds good.*) **But recently the chemical processes that have been thinning the ozone layer have been found to be less well understood than once thought.***destabilizing condition* We may have labeled hydrofluorocarbons as the chief cause incorrectly.*cost*

Readers now have not one but two reasons to see their self-interest in the problem: the problem itself, and also their ignorance of it. Common ground can describe a general misunderstanding:

The Crusades in the eleventh century are widely believed to have been motivated by religious zeal to restore the Holy Land to Christendom.*common ground* In fact, the motives were at least partly, if not largely, political.

Or survey current but perhaps flawed research:

Few sociological concepts have fallen out of favor as fast as Catholicism's alleged protective influence against suicide. Once one of sociology's basic beliefs, it has been called into question by a series of studies in both Europe and North America. . . . *common ground* However, certain studies still find an effect of religion . . .

Or it can point to a misunderstanding about the problem itself:

Education in the U.S. has focused on teaching children to think critically, to ask questions and test answers.*common ground* But the field of critical thinking has regularly been taken over by special programs based on fads and special interests. Until we recognize that there is no silver bullet way to teach critical thinking, it will not achieve what we wish it would.*problem*

Some inexperienced researchers skimp on common ground, thinking they can open their report as if they were picking up a class conversation where it left off. Their introductions are so sketchy that only others in the course would understand them:

In view of the controversy over Hofstadter's failure to respect the differences among math, music, and art, it was not surprising that the response to *The Embodied Mind* was stormy. What is less clear is what caused the controversy in the first place. I will argue that any account of the human mind must be interdisciplinary.

When you draft your introduction, imagine you are writing to another person who has read the same books and thought about some of the same issues, but does not know what happened in your particular class.

14.3 STATING YOUR PROBLEM

Once you establish common ground, you can disrupt it with a problem. As we said in chapter 4, the statement of a research problem has two parts:

- some *condition* of incomplete knowledge or understanding, and

- the *consequences* of not fully knowing or understanding.

You can state the condition directly:

Motodyne has no data showing which icons are self-explanatory.

Or you can imply it in an indirect question:

The real question is why these characters are always white and male.

This condition of ignorance or misunderstanding is part of a *full* research problem only when you then spell out a *consequence* as an answer to *So what?* You answer that question by stating a cost:

So what if you don't have that data?

Lacking such data, we cannot determine which icons to redesign.*cost*

Or you can transform the cost into a benefit:

> With such data, we can determine which icons to keep and
> which to redesign.*benefit*

This is not entirely a stylistic choice. Some research suggests that
readers are more motivated by a certain cost than by a possible
benefit.

That's the straightforward version of a problem; there are vari-
ations.

14.3.1 When Should You State Conditions Explicitly?

Occasionally, you tackle a problem so familiar that you can imply
it just by naming its condition. Such familiar conditions are
found in fields like mathematics and the natural sciences, in
which many research problems are widely known. Here again,
for example, is that abbreviated introduction to perhaps the most
significant article in the history of molecular biology, the one in
which Crick and Watson report their discovery of the double-helix
structure of DNA (this is substantially condensed):

> We wish to suggest a structure for the salt of deoxyribose nu-
> cleic acid (D.N.A.). This structure has novel features which are
> of considerable biological interest. A structure for nucleic acid
> has already been proposed by Pauling and Corey. They kindly
> made their manuscript available to us in advance of publication.
> Their model consists of three intertwined chains, with the phos-
> phates near the fibre axis, and the bases on the outside. In our
> opinion, this structure is unsatisfactory . . .

By saying that they will suggest a structure for DNA, Crick and
Watson implied that their readers did not know it. They did not
have to say that, because they knew readers were already keen
on the problem. (Note, though, that they do raise a problem to
solve by mentioning Pauling and Corey's *incorrect* model.)

In the natural sciences and most social sciences, researchers
typically address questions familiar to readers. Even so, readers
won't know what *particular* flaw in their knowledge your research

will address unless you tell them. In the humanities and some social sciences, researchers typically address questions that they alone have found or even invented, questions that readers find new and often surprising. In that case, you must explicitly describe the particular gap in knowledge or flawed understanding that you believe your readers can't resolve but should.

14.3.2 When Should You Spell Out Costs and Benefits?

To convince readers that your problem should matter to them, you must convince them to care about it because they will pay a cost if it is not resolved and gain benefits once it is. Sometimes you can describe tangible costs that your research helps your readers avoid (review pp. 64–67):

> Last year the River City Supervisors agreed that River City would benefit if it added the Bayside development project to its tax base. That argument, however, was based on little economic analysis. If the Board votes to annex Bayside without understanding what it will cost the city, **the Board risks worsening River City's already poor fiscal situation.** When the added burden of expanding city schools and bringing sewer and water service up to city code are included in the analysis, the annexation is less advantageous than the Board assumes.

This is the kind of problem found in applied research. The area of ignorance (no economic analysis) has tangible consequences in the world (unanticipated costs or benefits).

In pure research, you formulate the same kind of problem when you explain the cost not in dollars and cents, but as even greater flawed knowledge or misunderstanding, or alternatively, as the benefit of better understanding:

> Since 1972 American cities have annexed upscale neighborhoods to prop up tax bases, often bringing disappointing economic benefits. But that result could have been predicted had they done basic economic analysis. The annexation movement is a case study of how political decisions at the local level fail to use expert in-

formation. What is puzzling is why cities do not seek out those with expertise. **If we can discover why cities fail to rely on basic economic analyses, we might better understand why their decision-making fails so often in other areas as well.** This paper analyzes the decision-making process of three cities that annexed surrounding areas but ignored economic consequences.

14.3.3 Testing Conditions and Costs

In chapter 4 we suggested a way to test how clearly you have articulated the costs of not solving a problem: after the sentences that best state your condition of ignorance or misunderstanding, ask, *So what?* You have articulated your problem persuasively when what comes before the *So what?* plausibly elicits that question and when what follows plausibly answers it.

> Motodyne has no data showing which icons are self-explanatory. (*So what?*) With such data, it could determine which icons to retain and which to redesign.

> Stories about the Alamo in Mexican and U.S. versions differ in obvious ways, but U.S. versions from different eras also differ. (*So what?*) Well, hmmmmmm . . .

Answering that question is not just difficult; it can be exasperating, even dismaying. If you fall in love with stories about the Battle of the Alamo, you can pursue them to your heart's content, without having to justify your pursuit to anyone but yourself: *I just like knowing about it.* But for others to appreciate your research, you have to "sell" them on its significance. Otherwise, why should they spend time on it?

If you write a paper for a class, your teacher is obliged to read it, but when you address your research community, you have to convince them that your problem is *their* problem, that if they go on without knowing, say, how those stories about the Alamo have evolved, how Hollywood turned the story into myth, they will be neglecting something about their identity as North Americans.

Now, to be sure, some readers will ask again, *So what? I don't care about myth and history or our identity.* To which you can only shrug and think, *Wrong audience.* Successful researchers know how to find and solve interesting problems and how to convince readers that they have. But a skill no less important is knowing where to find readers who appreciate the kind of problem that you have solved.

If you are sure your readers know the consequences of your problem, you might decide not to spell them out. Crick and Watson did not specify either costs or benefits, because they knew that their readers would not understand genetics until they understood the structure of DNA. Had Crick and Watson stated those costs, they might have seemed both redundant and condescending.

If you are tackling your first research project, no reasonable teacher will expect you to articulate your problem in detail, because you probably do not yet know what other researchers think is significant. But you take a big step in that direction if you can state explicitly just *your own* incomplete knowledge or flawed understanding in a way that shows that *you* are committed to improving it. You take an even bigger step when you can explain why it is important to resolve that flawed understanding, when you can show that by understanding one thing better, you understand better something else much more important, *even if it is for you alone.*

14.4 STATING YOUR RESPONSE

Once you disrupt your readers' stable context with a problem, they will expect you to resolve it, either by explicitly stating the gist of your solution or by implicitly promising them that you will do so later on. They look for that response in the last few sentences of your introduction. You can state it in one of two ways.

14.4.1 State the Gist of Your Solution

You can state your solution explicitly. When you announce your main point in the introduction, you create a "point-first" paper

(even though that point appears as the *last* sentence of the introduction).

> As we have investigated environmental threats, our understand-
> ing of many chemical processes such as acid rain and the
> buildup of carbon dioxide has improved, allowing us to under-
> stand better their eventual effects on the biosphere.*common ground*
> (*Sounds good.*) But recently the chemical processes that have
> been thinning the ozone layer have been found to be less well
> understood than once thought.*condition* (*So what?*) We may have
> labeled hydrofluorocarbons as the chief cause incorrectly.*cost* **We
> have found that the bonding of carbon** . . . *gist of solution/main point*

14.4.2 Promise a Solution

Alternatively, you can put off stating your main point by stating
only where your paper is headed, thereby implying that you will
present your solution in your conclusion (review pp. 195–96).
This approach provides a "launching point" and creates a "point-
last" paper:

> As we have investigated environmental threats, our understand-
> ing . . . has improved. . . . But recently the chemical pro-
> cesses . . . [have proved to be] less well understood. . . . (*So
> what?*) We may have labeled hydrofluorocarbons as the chief
> cause incorrectly. (*Well, what* have *you found?*) **In this report, we
> describe a hitherto unexpected chemical bonding between** . . .

This introduction launches us into the body of the paper not with
a point or summary of its solution, but with a sentence that prom-
ises a solution to come.

The weakest form of a launching point is one that merely announces a topic:

> This study investigates the chemistry of ozone depletion.

If you have good reason to save your point for the end of your
paper, be sure that your launching point does more than just

announce a topic. It should suggest the conceptual outlines of
your solution or announce a plan (or both).

> There are many designs for hydroelectric turbine intakes and di-
> version screens, but on-site evaluation is not cost-effective. A
> more viable alternative is computer modeling. **To evaluate the hy-
> draulic efficiency of hydroelectric diversion screens, this study
> will evaluate three computer models, Quattro, AVOC, and Turbo-
> plex, to determine which is most cost-effective in reliability,
> speed, and ease of use.**

This kind of plan is common in social sciences, but less frequent
in the humanities, where many readers consider it ham-handed.

14.5 FAST OR SLOW?

A final decision is how quickly to raise your problem. That depends on how much your readers know. In the following, the
writer begins flat out, announcing a consensus among well-
informed engineers; then, in the second sentence, he briskly
disrupts that consensus:

> Fluid-film forces in squeeze-film dampers (SFDs) are usually ob-
> tained from the Reynolds equation of classical lubrication theory.
> However, the increasing size of rotation machinery requires the
> inclusion of fluid inertia effects in the design of SFDs.

We have no idea what that means, but we can see the pattern
clearly.

This next writer also addresses technical concepts but begins
with more familiar ones, implying readers who do not already
possess vast technical knowledge:

> A method of protecting migrating fish at hydroelectric power de-
> velopments is diversion by screening turbine intakes . . . [an-
> other 110 words explaining screens]. Since the efficiency of screens
> is determined by the interaction of fish behavior and hydraulic
> flow, screen design can be evaluated by determining its hydraulic
> performance . . . [40 more words explaining hydraulics]. This

study resulted in a better understanding of the hydraulic features of this technique, which may guide future designs.

When you open quickly, you imply an audience of peers; when you open slowly, you imply readers who know less than you. If your readers are knowledgeable and you open too slowly, you may sound as if *you* know too little. But if you open too quickly, you may seem inconsiderate of their needs.

14.6 ORGANIZING THE WHOLE INTRODUCTION

All this may seem formulaic, but when you master a rhetorical pattern, you have more than a formula for writing. You also have a tool for thinking. By forcing yourself to work through a full statement of your problem, you have to explore what your audience knows, what they don't, and, in particular, what they should.

By now you may feel overwhelmed with too many choices, but they all follow what is in fact a simple "grammar." A full introduction consists of just three elements:

Common Ground + Problem + Response

You don't need all three elements all the time:

• If the problem is well known, omit the common ground; begin with the condition of the problem.

• If the consequences of the problem are very well known, you can also omit them.

• If you want to show how you worked through the problem and solved it, state your main point in the conclusion; at the end of your introduction, frame your response as launching point.

Like all structural summaries, this one feels mechanical. But when you flesh this pattern out in a real paper, readers lose sight of the form and notice only the substance. In fact, the expected form helps them find the substance they are looking for. That form also encourages you to think harder than you might have.

14.7 CONCLUSIONS

Not every research paper has a section formally called *Conclusion*, but they all have a paragraph or two that serves as one. You may be happy to know that you can use the same elements that you used in your introduction for your conclusion. You just use them in reverse order.

14.7.1 Start with Your Main Point

• If you end your introduction with your main point, state it again at the beginning of your conclusion, but state it more fully. It should not simply repeat your introduction.

• If you end your introduction not with your main point but with a launching point, state your point at the beginning of your conclusion, and be sure to use the key terms you used at the end of your introduction.

14.7.2 Add a New Significance or Application

Once you state your claim, say why it's significant: paraphrase the consequences of your problem or point to a new significance not mentioned in your introduction. This new significance should be another answer to the question *So what?* in the introduction.

For example, in this next conclusion, the writer introduces for the first time an additional cost of the Supreme Court's decision on military death sentences: the military may have to change the culture of its thinking.

> In light of recent Supreme Court decisions rejecting mandatory capital punishment, then, the mandatory death provision for treason is apparently unconstitutional and must therefore be revised by Congress. More significantly, **though, if the Universal Code of Military Justice is changed, it will challenge a fundamental value of military culture: ultimate betrayal mandates the ultimate penalty. Congress will then have to deal with the military's universal sense of what is just.**

The writer could have used that implication in his introduction, as a potential cost resulting from new Supreme Court decisions, but he may have felt that it was too volatile to raise early.

As you write your conclusion, take care not to broaden a possible significance so much that it seems to be your main point. You can be clear about its role by introducing it almost "by the way," as an additional, *possible* practical implication of your solution.

14.7.3 Add a Call for More Research

Just as you can survey research already done in your common ground, you can also call for more research still to do at the end of your conclusion:

> These differences between novice and expert diagnosticians clearly define the starting and ending points in their maturation and development. We know how novices and experts think differently. **What we do not understand is which elements in the social experience of novices contribute to that development and how. In particular, we need longitudinal studies on how mentoring and coaching affect outcomes, whether active explanation and critique help novices become skilled diagnosticians more quickly.**

When you conclude by pointing out what remains to be done, you show your readers that you haven't had the last word on your problem, that there is still more to say. That keeps the conversation alive. Those who pursue your suggestion will review your work, respond to it, and move beyond it. So before you write your last words, imagine someone fascinated by your work who wants to follow up on it: What would you suggest they do? What more would you like to know? After all, that may have been one of your strategies in finding a problem of your own (look again at pp. 68–70).

Many writers find the very first sentence or two especially difficult to write, and so they fall into clichés:

- Don't start by citing a dictionary entry: Webster's *defines* ethics *as* . . . If a word is important enough to define in a report, it is too complex for a dictionary definition.

- Don't start grandly: *The most profound philosophers have for centuries wrestled with the important question of* . . . If your subject is grand, it will speak its own importance.

- Don't repeat the language of your assignment. If you are struggling to start, prime your pump with a paraphrase, but when you revise, drop it.

Here are three choices for your first sentence or two.

OPEN WITH A STRIKING QUOTATION
Do this only if its language is like the language in the rest of your introduction:

> "From the sheer sensuous beauty of a genuine Jan van Eyck there emanates a strange fascination not unlike that which we experience when permitting ourselves to be hypnotized by precious stones."
>
> Edwin Panofsky, who had a way with words, suggests here something magical in Jan van Eyck's works. His images hold a fascination . . .

OPEN WITH A STRIKING FACT

> Those who think that tax cuts for the rich stimulate the economy should contemplate the fact that the top 1 percent of Americans own as much wealth as everyone in the bottom 40 percent.

OPEN WITH A RELEVANT ANECDOTE

Again, do this only if its language or content connects to your topic and if it vividly illustrates an aspect of your problem. The following paper addressed the economics of school segregation:

> This year Tawnya Jones begins junior high in Doughton, Georgia. Though her classmates are mostly African American like herself, her school system is considered legally racially integrated. But except for a few poor whites and Hispanic students, Tawnya's school still resembles the segregated and economically depressed one that her mother entered in 1962. . . .

When you open with any of these devices, be sure to use language that leads to your context, your problem, and a focused statement of its solution.

CLOSE WITH AN ECHO

You bring your conclusion to a graceful, even literary close by echoing your opening fact, anecdote, or quotation with another at the end. Here, for example, is an introduction that begins with a quotation, an epigraph that highlights the themes of spiritualism and modernity. The writer echoes those themes with a parallel quotation at the end of her conclusion (note, too, how the title pulls together key themes):

> Flannery O'Connor and the Spiritual Foundations of Racism:
> Suffering as Southern Redemption in the Modern World
> *"I write the way I do because . . . I am a Catholic peculiarly*
> *possessed of the modern consciousness."*
>
> Although Flannery O'Connor's stories give us insights into southern culture, some have said her attitude toward race was the product of "an imperfectly developed sensibility" and that "large social issues as such were never the subject of her writing." But that criticism ignores . . .

Here is the conclusion:

> Thus we see that those who claim that O'Connor ignored racism fail to see that she understood racism as a deeper crisis of faith,

as a failure to recognize the healing knowledge of suffering, insights that put her among a few southern writers who saw the modern world as spiritually bankrupt. Seen in this light, a rereading of her private correspondence might reveal . . . **As she said in one letter (May 4, 1955), "What I had in mind to suggest [was] . . . the redemptive quality of the Negro's suffering for us all. . . . I meant [a character in the story to suggest] in an almost physical way . . . the mystery of existence."** *conclusion*

This echoing device may seem a bit literary, but it is not at all uncommon.

CHAPTER FIFTEEN

Communicating Evidence Visually

This chapter focuses on the issues involved in presenting quantitative data in tables, charts, and graphs clearly, usefully, and fairly.

As we've said in many ways, readers assess a claim by the strength of the argument supporting it, particularly the soundness of its logic and the quality of its evidence. Since readers rightly insist on evidence, particularly new evidence, you have to be sure that they understand yours easily and see its relevance to the claim you intend it to support. That is especially so when the evidence consists of complex quantitative data whose impact can be strengthened or weakened by how you present them. So if you have based your report on lots of complex data, particularly quantitative data, you should now focus on how clearly you have presented them and revise those tables and figures that do not clearly and persuasively connect your reports of evidence to your claims.

Some reports of quantitative data are just as clear verbally as visually:

In 1996, on average, men earned $32,144 a year, women $23,710, a difference of $8,434.

TABLE 15.1
Male and Female Salaries, 1996

Men	$32,144
Women	$23,710
Difference	$ 8,434

But when the numbers are more complex, readers need a more systematic presentation, first simply to absorb them, then to analyze and understand them. For example, here is a paragraph of data too complex to remember easily.

In 1970 almost nine out of ten families had two parents—85 percent. But in 1980 that number declined to 77 percent, then to 73 percent in 1990, and to 68 percent in 2000. The number of one-parent families rose, particularly families headed by just a mother. In 1970 just 11 percent of families were headed by a single mother. In 1980 that number rose to 18 percent, in 1990 to 22 percent, and to 23 percent in 2000. Single fathers headed just 1 percent of the families in 1970, 2 percent in 1980, 3 percent in 1990, and 4 percent in 2000. Families with no adult in the home have remained stable at 3–4 percent from 1970–2000.

Those numbers would be far more accessible in a table:

TABLE 15.2: Changes in Family Structure, 1970–2000

Family type	Percent of total families			
	1970	1980	1990	2000
2 parents	85	77	73	68
mother	11	18	22	23
father	1	2	3	4
no adult	3	4	3	4

Or as a bar chart:

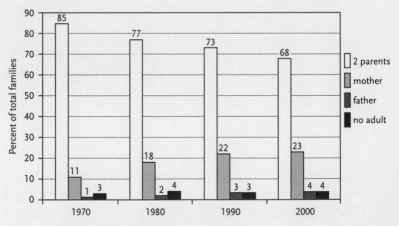

Figure 15.1: Changes in Family Structure, 1970–2000

Or as a line graph:

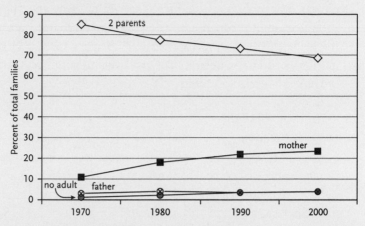

Figure 15.2: Changes in Family Structure, 1970–2000

Readers can get the same data from each of those more visually oriented representations, but they will experience different rhetorical effects:

- The table of numbers feels precise and objective. It does not impose on us any predigested outcome. It lets us compare the numbers systematically and come to our own conclusion.

- The bar chart gives us less exact information (though that is compensated for by adding the numbers above the bars). But it visually communicates the gist of its point quickly. It helps us make individual comparisons.

- The line graph also gives us less exact information but offers an even more striking image of a story. It helps us see trends easily.

These considerations underscore a point we made in chapter 9: You must report evidence in a way that is clear, appropriate, and fair, but every choice unavoidably "spins" your evidence, giving it a particular rhetorical effect. Your task is to create the effect that best serves your intention but does not mislead readers, a choice that influences how readers respond not only to your data, but to you. So you have to choose your format thoughtfully. You must choose among tables, charts, and graphs first to construct a clear report of your evidence, second to create the appropriate rhetorical effect, and finally to avoid depictions of data that are misleading, or at worst deceitful. Too many researchers these days subordinate a respect for truth to their desire for striking but deceptive visual effects.

15.1 VISUAL OR VERBAL?

Different readers prefer data to be reported verbally or visually, so you first have to understand your readers' needs and expectations. Readers who are word-oriented (most often in the humanities) are less comfortable with complex visual representations of data than are those in the social and natural sciences. So if you have relatively simple data to report to humanists, report them verbally, as in the paragraph about comparative earnings.

15.2 TABLES VS. FIGURES

If, however, your data are too complex to report in words, report them in visual form, in a table or figure, such as a bar chart or line graph. (There are many more sophisticated choices, but we can offer only the basics of representing data. After you master these basics, see p. 323 for references to more advanced treatments.) Here are two general guidelines:

- Choose a table if your readers are likely to want very precise numbers and you don't want to impose on your data a visual image implying the point you want them to support.

- Choose a figure if your readers are less interested in precise details than in a general point, and you want to reinforce your point with a strong image.

Yet there are qualifications. For example, if you had a vast amount of quantitative data about lots of diseases and injuries for all fifty states, categorized by gender, race, ethnicity, urban vs. rural, and so on, no single figure could represent such complexity; you would have to use at least one table. But when you can choose between a table and a figure, balance what you intend to communicate against what readers are likely to value more: precision or a visual impact.

15.3 CONSTRUCTING TABLES

If you choose to present your data as a table, observe the following principles (the first two apply to graphs and charts, as well).

1. Introduce your data with a sentence that explicitly tells the reader what to see in them. Then give the table, graph, or chart a title that explicitly names its purpose.

2. Organize your table, bar chart, or line graph in a way that anticipates how your readers will use it, and highlight those data most relevant to the claim you want the data to support.

For example, on first reading, it's hard to see how these next data relate to the claim they seem intended to support because we have to do a lot of calculating:

> Though the United States has had unprecedented economic growth in the last twenty-five years that has benefited some, most Americans have lost ground.

TABLE 15.3: Income

	1977	1999
Bottom 20%	$10,000	$8,800
Second 20%	$22,100	$20,000
Third 20%	$32,400	$31,400
Fourth 20%	$42,600	$45,100
Top 20%	$74,000	$102,300
Top 1%	$234,700	$515,600

We would understand the relevance of those data more quickly and clearly with four changes: (1) a prior sentence interpreting them, (2) an informative title specifying the topic, (3) the key comparisons calculated, and (4) their results highlighted. (In tables, negative numbers are typically represented in parentheses.)

Though the United States has had unprecedented economic growth in the last twenty-five years that has benefited some, most Americans have lost ground.*claim* **Between 1977 and 1999, the top 20 percent of wage earners increased their income by more than 38 percent, and the top 1 percent more than doubled theirs, but the bottom 60 percent of the population earned less in 1999 than they did in 1977.***reason*

TABLE 15.4: Changes in After-Tax Annual Income 1977–1999 (by quintile)

	1977 $	1999 $	± % changes
Bottom 60%	$21,500	$20,000	(7.0)
Bottom 20%	$10,000	$8,800	(12.0)
Second 20%	$22,100	$20,000	(9.5)
Third 20%	$32,400	$31,400	(3.1)
Fourth 20%	$42,600	$45,100	5.9
Top 20%	$74,000	$102,300	38.3
Top 1%	$234,700	$515,600	119.7

Never force your readers to figure out on their own what you want them to see in a table or figure. In an introductory sentence, tell them what to see, reinforce that with an informative title, and then, if you can, visually highlight key data.

To those first two principles for constructing tables in particular add these six:

3. Down the left-hand side of the table, list the items whose numbers you are presenting to the right.

4. Across the top, list the categories of data. If your table represents a sequence of months, years, and so on, put them across the top.

5. Group and order the items running down the left side and across the top so that what goes together conceptually is grouped together visually; present everything in an order that helps readers find what you want them to look for quickly and

reliably. Choose an alphabetical order only if there are a lot
of items and you have no particular point to draw from the
data.

6. Don't clutter a table with horizontal and vertical lines separating all rows and columns. If there are five to seven rows,
use *faint* separating lines horizontally; for eight to twelve
rows, put a small space or heavier line between every four
rows; for very large tables, use faint gray scale for rows at
regular intervals (every other row, every fifth row, etc.).

7. Make your numbers relevant to your readers' needs by round-
ing to eliminate irrelevant differences. The numbers 2,123,000
and 2,124,000 may be irrelevantly precise if no decision or
judgment will turn on a difference of 1,000. In most cases,
you would help readers by representing both as 2.1 million.

For example, suppose you wanted table 15.5 to show that
English-speaking nations have reduced unemployment most in
recent years. How easy is it to find the relevant data?

TABLE 15.5: Unemployment Rates of Major Industrial Nations

	1990	2001	Change
Australia	6.7	6.5	(.2)
Canada	7.7	5.9	(1.8)
France	9.1	8.8	(.3)
Germany	5.0	8.1	3.1
Italy	7.0	9.9	2.9
Japan	2.1	4.8	2.7
Sweden	1.8	5.1	3.3
UK	6.9	5.1	(1.8)
USA	5.6	4.2	(1.4)

Table 15.6 has a title that makes its point more clearly, but it also gets rid of clutter, groups nations by language, orders each group by degree of change, and highlights the four relevant data points.

TABLE 15.6: Changes in Unemployment Rates
English-speaking vs. Non-English-speaking Nations

	1990	2001	Change
Australia	6.7	6.5	(.2)
USA	5.6	4.2	(1.4)
Canada	7.7	5.9	(1.8)
UK	6.9	5.1	(1.8)
France	9.1	8.8	(.3)
Japan	2.1	4.8	2.7
Italy	7.0	9.9	2.9
Germany	5.0	8.1	3.1
Sweden	1.8	5.1	3.3

15.4 CONSTRUCTING FIGURES

Choose a figure over a table if precise numbers are less important than readers' getting an image of the story in your data. Which of these images tells its story more powerfully?

TABLE 15.7: Rise in Public and Private Spending on Health (in Billions), 1960–1999

	1960	1970	1980	1990	1999
Private	20.1	45.5	141.0	413.2	662.1
Public	6.6	27.6	104.8	282.4	548.1

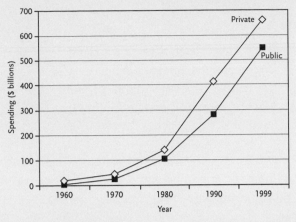

Figure 15.3: Rise in Public and Private Spending on Health (in Billions), 1960–1999

We now need two technical terms to explain how charts and graphs work and how to construct clear and fair ones: they are *axis* and *variable*.

AXIS. Graphs and charts have two formal elements, a vertical Y-axis and a horizontal X-axis.

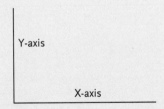

Figure 15.4

VARIABLES. Graphs and charts represent two kinds of content, both called *variables*. The two kinds are *independent* and *dependent*.

- *Independent* variables are the established criteria of measurement: the continuous numbers referring to weight, profits, temperature, volume, decibels, speed, and so on.

—In bar charts, the independent variable is plotted on the vertical Y-axis. Measures of time are usually an exception. They are typically plotted on the X-axis.

—In line graphs, one independent variable is plotted on the vertical Y-axis, a second independent variable is plotted on the horizontal X-axis, especially when that second variable is time.

- *Dependent* variables are the discrete things being measured: companies, voter turnout, people, cars, dollars, planets, explosions, injuries, cases of cancer, popularity of a work of art.

  —In a bar chart, dependent variables are plotted along the horizontal X-axis.

  —In a line graph, they are plotted at the intersections of the independent variables: speed and injuries, sales and profits, height and weight, and so on.

There is a general principle in choosing between a graph and a chart:

- Choose a vertical bar chart to represent static situations, where entities (the dependent variables) are measured at a moment in time:

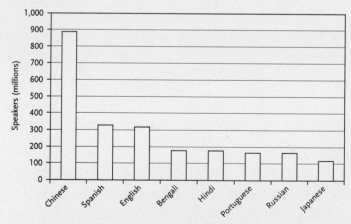

Figure 15.5: Languages Spoken by More than 100 Million Speakers

- Choose a line graph to represent dynamic relationships—movement through time, or correlations between changing measures, such as height and weight, speed and distance:

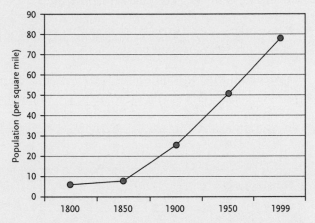

Figure 15.6: Increases in Population Density in the United States, 1800–1999

Most readers can see the contour of growth in both a line graph and bar chart, but the line graph represents the image of growth more clearly and economically.

There are some general principles to keep in mind in constructing both bar charts and line graphs:

- As you do with tables, introduce all figures with a sentence that tells the reader the point of the data and create a title that reinforces the point.

- Use "tick" marks on the vertical Y-axis to help readers see the point of measure; if the tick marks are finely graded, boldface every fifth one.

- Use faint grid lines to help readers estimate numbers (run the grid lines *behind* the bars).

15.4.1 Line Graphs

There are three principles in constructing line graphs in particular:

- Keep the image as uncluttered as possible. If you are plotting more than four dependent variables (the things being measured), you risk confusing your readers.

- If you cannot divide a complex graph into two graphs, then clearly distinguish the lines for each element; if you can, label the lines rather than using a legend (even though the added labels further complicate the image).

- Help readers see clearly the data points on the line. Put a dot at each relevant data point.

Compare figure 15.7 and figure 15.8. Which is easier to understand? What does that graph do to help you understand its data?

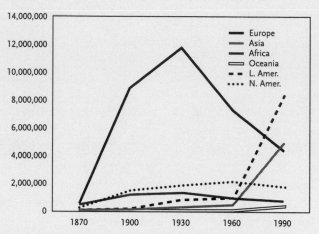

Figure 15.7: Foreign-born Residents in the United States, 1870–1990

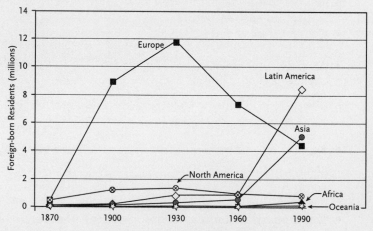

Figure 15.8: Foreign-born Residents in the United States, 1870–1990

15.4.2 Bar Charts

As we've said, choose a bar chart to compare discrete dependent variables at a single moment in time. It is possible to represent time for multiple entities on a bar chart, but the image becomes very complicated, making it difficult for a reader to see relevant differences. In effect, you get a series of little bar charts imposed on a single big one. Compare this bar chart to figure 15.8:

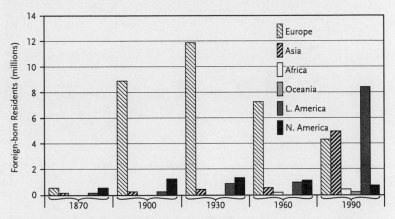

Figure 15.9: Foreign-born Residents in the United States, 1870–1990

If your X-axis does not represent time, you are generally free to order the bars along the X-axis as you wish, but there are some principles for doing that:

- Group the bars into related sets whenever possible.

- Arrange the bars so that they give an image of order.

- Highlight a bar if it is a relevant point of comparison for the others.

- Keep visual contrasts simple: black, white, and one or two shades of gray. If possible, avoid cross-hatching, stripes, and so on (impossible if you try to chart too many cases).

- If necessary, include numbers above the bars to give readers more precision.

Contrast figure 15.10 with 15.11. Assume that the data are intended to support this claim:

Most of the desert area in the world is concentrated in North Africa and the Middle East:

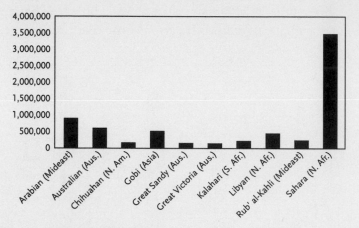

Figure 15.10: World's Ten Largest Deserts

Figure 15.10 is organized alphabetically, but that does not help readers find the data that support the claim. We see no numbers associated with the bars. No grid lines help us connect numbers to the Y-axis. And the Y-axis has no tick marks. In contrast, figure 15.11 is organized into a coherent picture to support that claim:

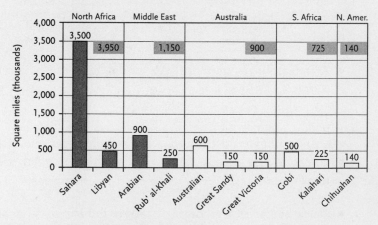

Figure 15.11: World Distribution of Large Deserts

15.4.3 Stacked Bar Charts

Stacked bars are a variation on side-by-side bars. Stacked bars divide the bar into its relative proportions of 100 percent of some other variable. They can be difficult to process because they force readers to make comparisons and gauge proportions by eye alone. In figure 15.12 which world region has the fastest growth in nuclear power? Can you even see the tiny section for the Mideast?

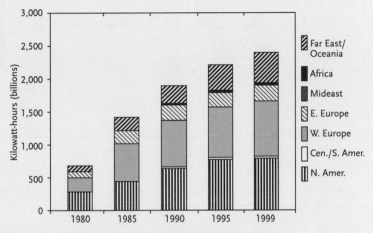

Figure 15.12: World Generation of Nuclear Energy, 1980–1999

If you insist on using stacked bars, help readers by following these principles:

- Arrange the segments in some principled order, from bottom to top. If feasible, put the largest elements at the bottom, smaller ones on top, and use the darkest color at the bottom, lightest at the top.

- Use numbers and connecting lines to clarify proportions.

- Don't bother to include cases whose numbers are so small that they are dwarfed by larger ones.

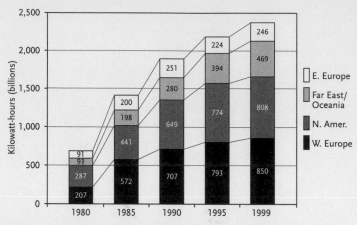

Figure 15.13: Largest Generators of Nuclear Energy, 1980–1999

15.4.4 Horizontal Bar Charts

Some researchers tip a vertical bar chart on its side to create a horizontal bar chart. The Y-axis is now horizontal and the X-axis is vertical. About the only advantage of a horizontal bar chart over a vertical one is typographical: it lets you get the whole name of an item next to a bar:

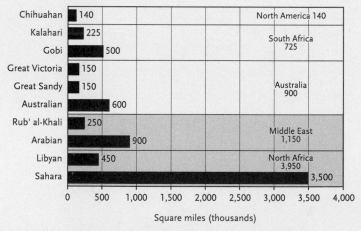

Figure 15.14: World Distribution of Large Deserts

15.4.5 Centrally Divided Horizontal Bar Charts

A variation on a horizontal bar chart is a centrally divided horizontal bar chart. It puts two dependent variables on either side of a center line and then displays a number of independent variables. The same data can be represented in a side-by-side vertical bar chart, but it is typographically more difficult to do:

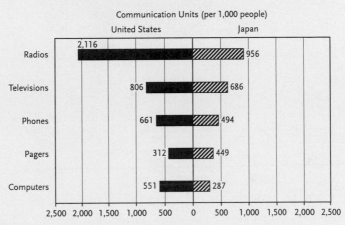

Figure 15.15: Electronic Communications, Year 2000

15.4.6 Pie Charts

Pie charts are favorites of newspapers and annual business reports but are often considered a bit amateurish for academic research. At best, they allow readers to see crude proportions among a few elements that constitute 100 percent of a whole. They are hard to read when they have more than four or five segments, particularly when the segments are thin and readers have to look at a key to match the patterns in the segments with categories. When readers try to judge the relative size of segments, they are likely to be wrong.

For example, using this pie chart, how easily could you compare the number of Japanese speakers to Hindi? What is the ratio of Chinese speakers to Portuguese?

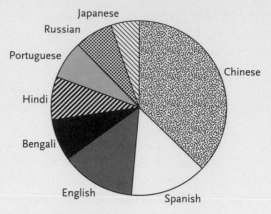

Figure 15.16: Languages with More than 100 Million Speakers

Now answer the same questions with a much easier bar chart:

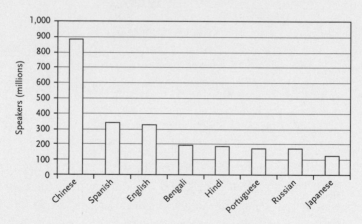

Figure 15.17: Languages with More than 100 Million Speakers

If you insist on using a pie chart, here are some principles:

- Arrange the segments in an order meaningful to your readers, beginning at 12 o'clock and moving clockwise. If you have no better order, arrange the segments from largest to smallest.

- If one segment is significant, emphasize it by coloring it or breaking it out from the rest.

- Don't use a legend; label the segments directly.

15.4.7 Avoid Three-Dimensional Graphics

Almost every office software package now includes software that allows you to create ornate graphics, with multiple colors and shapes in three dimensions. Our advice is simple: Don't. Only rarely are data complex enough to require a three-dimensional representation. Most often, the third dimension is purely decorative, which is to say distracting from its point. For example, you can create this "field of cones" with only a few clicks in the most popular spreadsheet program:

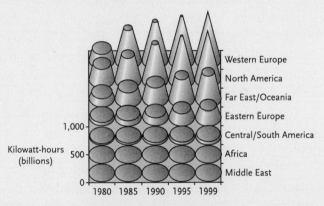

Figure 15.18: World Generation of Nuclear Energy, 1980–1999

But there is one kind of graphic even worse than the overly decorative options in software packages: iconic graphics that use drawings as elements in a chart. You've seen them in *USA Today* and other popular publications: a bar chart with football players of different heights representing annual salaries or steaks of different sizes representing annual beef consumption. These may seem flashy or cute, but readers of research reports don't want flashy or cute: they want useful information usefully presented.

You would lose credibility with your readers if you tried to represent oil imports like this:

Country A

Country B

Country C

All Others

Figure 15.19: Oil Imports, 1980–1999

15.5 VISUAL COMMUNICATION AND ETHICS

When you select a visual for its impact, remember that your rhetorical decision has an ethical dimension. Whenever you present data visually, you have to balance your rhetorical goals and your responsibility not just to the facts but to the fairness of their appearance. Tables, charts, and graphs always seem objective, and so they can fool inexperienced readers. But they will make experienced readers suspicious if you seem to distort the image to serve your story.

Unfortunately, it is sometimes difficult to distinguish effective rhetorical impact from unfair manipulation. For example, compare the two charts in figure 15.20. The data in the two are identical, but look at the slope of the bars:

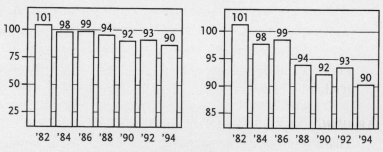

Figure 15.20: Capitol City Pollution Index, 1982–1994

On the left, the slope represents changes in data points more accurately, because the scale begins at 0, making the difference between readings of 101 in 1990 and 90 in 2002 look relatively small. On the right, the slope is much sharper, because the scale begins at 80, magnifying changes between 1990 and 2002. As a result, the chart on the right suggests more improvement, a story that might mislead some readers and that might even be considered dishonest by others.

The distortion in figure 15.20 is mitigated by the fact that the bars are clearly labeled with precise values. But a writer who truncates the vertical axis of a graph to make a slope seem sharper, as in the next figure, may cross the line of honesty, because to the viewer the slope of a graph is always the predominant image:

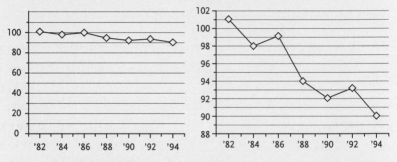

Figure 15.21: Capitol City Pollution Index, 1982–1994

On the other hand, it is not always easy to distinguish what is "objective" from what is "ethical." Suppose you are an environmental scientist and you know that any expert would consider these seemingly small decreases to be highly significant. If you were certain that your statistically unsophisticated readers would dismiss the visually slight differences on the left as meaningless, then that larger visual difference on the right would more accurately communicate the real scientific significance. In that case, it's harder to decide which graph is more honest.

15.6 USING GRAPHICS AS AN AID TO THINKING

Like any formal device, graphical representations of data can help you to see things in new ways, to see trends, discover new relationships, recognize the significance of a particular set of data. Some mathematical types can see all of this by just looking at the numbers. But the rest of us need visual representations to see what there is to see in a set of numbers. So don't just think of graphics as an appealing way to show readers lots of data. Try out as many combinations of data and kinds of representations as your imagination can dream up and your software can execute. You can never tell what insight you will have after looking at those pictures. We display data not just to make them clear, but to help us see them in new and striking ways. But that's something well beyond the limits of what we can offer here.

CHAPTER SIXTEEN

Revising Style: Telling Your Story Clearly

So far, we have urged you to focus more on the content and organization of your report than on its sentences. But readers have to understand your sentences to understand your argument. When you approach a final draft and are ready to revise for style, the steps in this chapter will help you do it efficiently.

If readers are to accept your claim, they must be able to follow your argument, and to do that, they have to understand the sentences that express it. So once you've revised the body of your paper to make a clear and cogent argument and you've redone your introduction so that it frames your argument in a way that makes it seem worth reading, you will have to focus on the details, making your sentences as clear as the complexity of your ideas allows.

But again, you face a complicated problem: you can't predict how readers will judge your style just by reading what you've written, because you already know what you want your sentences to mean.

16.1 JUDGING STYLE

If you had to read a long report written in the style of one of the following three examples, which would you choose?

> 1a. Too precise a specification of information-processing requirements incurs a risk of a decision-maker's over- or underestimation, resulting in the inefficient use of costly resources. Too little precision in specifying needed processing capacity gives no indication with respect to the means for the procurement of needed resources.

1b. A person who makes decisions may specify what he needs to process information. He may do so too precisely. He may over- or underestimate the resources that he needs. When he does that, he may use costly resources inefficiently. He may also fail to be precise enough. He may not indicate how others should procure those resources.

1c. When a decision-maker specifies too precisely the resources he needs to process information, he may over- or underestimate them and thereby use costly resources inefficiently. But if he is not precise enough, he may not indicate how those resources should be procured.

Few readers choose (1a); some choose (1b); most choose (1c). Example (1a) sounds like a machine speaking to a machine (it actually appeared in a respectable journal). Example (1b) is clearer but almost simpleminded, like an adult speaking slowly to a child. Example (1c) is clearer than (1a), but not condescending like (1b); it sounds more like a colleague speaking to a colleague. One of the worst problems in academic writing today is that too many researchers write like (1a).

Some would disagree, claiming that heavy thinking demands heavy writing, that some ideas are so intrinsically complex that when writers try to make them clear, they oversimplify, sacrificing nuances and complexity of thought. For them, if readers can't understand, well, that's their problem.

Perhaps. Everyone who reads the philosophers Immanuel Kant or Friedrich Hegel struggles with their style, at least at first. But most serious readers concede that understanding what they have to say is worth the effort. The problem is, few of us are a Kant or Hegel. For most of us most of the time, such complex writing is more likely to reflect sloppy thinking than the irreducible difficulty in the ideas of genius. Even when complex thinking warrants a complex style (less often than we think), every sentence profits from a second look; no sentence (including any of ours) is above revision (in fact, Kant and Hegel could have profited from good editors of their own).

Of course, some writers go too far in avoiding a complex style, using only short, simple sentences like those in (1b) above. But we assume that most of you reading this chapter do not have that problem, and that you need little help with your spelling and grammar. (If you think you need such help, ask your teacher to recommend a handbook and tell you where your school's writing tutors hang out.) We address here the problem of a style that is unnecessarily complex, too "academic," more difficult than it has to be.

This problem especially afflicts those just starting advanced work because they are hit by double trouble. First, when they have to write about complex ideas that test their comprehension, their style breaks down. Second, they often compound the problem when they take as a model those writers whose prose is terminally dense, because they think that a complex style bespeaks academic success. They are wrong. Convoluted, indirect, impersonal prose does not represent what truly expert writers can write, but what thoughtless writers are able to get away with.

16.2 A FIRST PRINCIPLE: STORIES AND GRAMMAR

When you chose among the three examples above, you probably evaluated each one using words like *clear* or *unclear, concise* or *wordy, direct* or *indirect*. But notice that those words really refer to how you *felt* about the sentences, to your *impressions* of them. If you said that (1a) was *dense,* you were really saying that *you* had a hard time getting through it; if you said (1c) was clear, you were saying that *you* found it easy to understand. There is nothing wrong with impressionistic language, but it does not explain *what on the page makes you feel as you do*. To do that, you need a way of talking about sentence style that lets you connect your impressions to what causes them.

The principles that distinguish the felt complexity of (1a) from the mature clarity of (1c) are few and simple. Those principles will direct your attention to only two parts of your sentences: to the first six or seven words and to the last four or five. If you can get those few words straight, the rest of the sentence will usually take care of itself. To understand these principles, though, you must first understand five grammatical terms: *subject, verb, noun, preposition,* and *clause.* (If you haven't thought about those terms for a while, refresh your memory before you go on.)

Keep in mind that our advice applies to revision. In chapter 12 we urged you to draft quickly, to get something down on paper *before* you start concentrating on details of sentence structure, punctuation, or spelling. If you apply the advice here about revising *as you draft,* you may tie yourself in knots. Save your concern for style until you have something to revise.

16.2.1 Subjects and Characters

The first principle may remind you of something you learned in grammar school, but it is in fact more complicated. At the heart of every sentence are its subject and verb. At the heart of every story are its characters and actions. In grammar school you probably learned that subjects are characters (called "doers"). But that is not always true, because subjects can refer to things other than characters. Compare these two sentences (the whole subject in each clause is underlined):

2a. <u>Locke</u> frequently repeated himself because <u>he</u> did not trust words to name things accurately.

2b. <u>The reason for Locke's frequent repetition</u> lies in his distrust of the accuracy of the naming power of words.

The subjects in (2a) fit that grammar-school definition: the subjects—*Locke* and *he*—are doers. But the subject of (2b)—*The reason for Locke's frequent repetition*—certainly does not, because *reason* is not a character.

We can see the same difference between these two (whole subjects are underlined):

3a. If <u>rain forests</u> are continuously stripped to serve short-term economic interests, <u>the entire biosphere</u> may be damaged.

3b. <u>The continuous stripping of rain forests in the service of short-term economic interests</u> could result in damage to the entire biosphere.

In the clearer version, (3a), look at the first few words of each clause:

3a. If <u>rain forests</u>_{*subject*} are continuously stripped_{*verb*} . . . <u>the entire biosphere</u>_{*subject*} may be damaged._{*verb*}

Those subjects name the main characters in a few short, concrete words: *rain forests* and *the entire biosphere*. Compare this:

3b. <u>The continuous stripping of rain forests in the service of short-term economic interests</u>_{*subject*} could result_{*verb*} in damage to the entire biosphere.

In (3b) the subject does not express a concrete character in a few concise words but rather an action in a long complex phrase: *The continuous **stripping** of rain forests in the service of short-term economic interests.*

If we can agree that (2a) and (3a) are clearer than (2b) and (3b), then we can see why grammar-school definitions may be bad language theory but good advice about writing. The first principle of clear writing is this:

Readers will judge your sentences to be clear and readable to the degree that you can make the subjects of your verbs name the main characters in your story. In particular, make your subjects short, specific, and concrete.

16.2.2 Verbs, Actions, and "Nominalizations"

There is a second key difference between clear and unclear prose: it depends on how writers express the crucial *actions* in their stories—as verbs or as nouns. For example, look again at the pairs of sentences (2) and (3) below. (Words naming actions are boldfaced; actions that are verbs are underlined; actions that are nouns are double-underlined.)

> 2a. Locke frequently **repeated** himself because he did not **trust** words to **name** things accurately.

> 2b. The reason for Locke's frequent **repetition** lies in his **distrust** of the accuracy of the **naming** power of words.

> 3a. If rain forests are continuously **stripped** to **serve** short-term economic interests, the entire biosphere may be **damaged**.

> 3b. The continuous **stripping** of rain forests in the **service** of short-term economic interests could result in **damage** to the entire biosphere.

Sentences (2a) and (3a) are clearer than (2b) and (3b) because their subjects are characters, but also because their actions are expressed not as nouns but as verbs: *repeated* vs. *repetition,* the verb *trust* vs. the noun *distrust;* the verb *name* vs. *naming power, stripped* vs. *stripping, serve* vs. *service,* the verb *damaged* vs. the noun *damage.*

Moreover, when you express actions not with abstract nouns but with verbs, you get rid of the clutter of prepositions. Look at the prepositions (boldfaced) in (4a) that (4b) doesn't need:

> 4a. Our development and standardization **of** an index **for** the measurement **of** thought disorders has made possible quantification **of** response **as** a function **of** treatment differences.

> 4b. Now that we have developed and standardized an index to measure thought disorders, we can quantify how patients respond to different treatments.

You're forced into using lots of prepositions—in this case, four *of*s, one *as,* and one *for*—when you turn verbs into nouns: *develop* → *development, standardize* → *standardization, measure* → *measurement, quantify* → *quantification, respond* → *response.*

There is a technical term for turning a verb (or an adjective) into a noun: we *nominalize* it. (When we nominalize the verb *nominalize,* we create the nominalization *nominalization.*) Most nominalizations end with endings such as *-tion, -ness, -ment, -ence, -ity*. But some are spelled like the verb:

Verb	Nominalization	Adjective	Nominalization
decide	decision	precise	precision
fail	failure	frequent	frequency
resist	resistance	intelligent	intelligence
delay	delay	specific	specificity

When you nominalize adjectives and verbs, you change your sentences in two other ways:

- You have to add verbs, which will always be less specific than the ones you could have used.

- You are likely to make the characters in your story modifiers of nouns or to drop them from a sentence altogether.

So here are two basic principles of a clear style:

- Make your central characters the subjects of your verbs; keep those subjects short, concrete, and specific.

- Use verbs to express crucial actions.

16.2.3 Diagnosis and Revision

From these principles of reading, we can offer two principles of writing, one for diagnosis and one for revision:

To diagnose:

1. Draw a line under the first six or seven words of every clause, whether main or subordinate, at the beginning, middle, or end of a sentence.

2. If those first six or seven words are subjects that are not characters but abstractions, and if the verb is a general one like *have, do, make, be,* and so on, that sentence is one you should probably revise.

To revise:

1. First, locate the characters you want to tell a story about. If you can't find any, decide who ought to be the main characters.

2. Next, look for what those characters are doing. If their action is in a nominalization, change it into a verb (i.e., "denominalize" it) and make the character its subject.

Remember that you may have to recast your sentence around some version of *If X, then Y; Because X . . . , Y; Although X, Y; When X, then Y.*

That's the simple version. Now we must make it a bit more complex.

16.2.4 Who or What Can Be a Character?

You may have been surprised when we called *rain forests* and *the entire biosphere* "characters," because one usually thinks of characters as flesh-and-blood. And, in fact, most readers prefer prose in which characters are flesh-and-blood people.

But we can also tell stories whose characters are abstractions. In your kind of research, you may have to tell a story about *demographic changes, social mobility, unemployment, isotherms, magnetism,* or *gene pools.* Sometimes you have a choice: your paper in economics might tell a story about people, such as *consumer, the Federal Reserve Board,* and *Congress,* or about abstractions associated with them, such as *savings, fiscal policy,* and *legislation.*

5a. When consumers save more, the Federal Reserve adopts a fiscal policy that influences how Congress legislates on taxes.

5b. Increased savings result in a Federal Reserve policy that influences congressional tax legislation.

In this sense, a character is any entity, real or abstract, that you focus on through several sentences, often making it the subject of those sentences. A passage might be about people or about the abstractions associated with them: *bankers* vs. *fiscal policy, savers* vs. *microeconomics,* or *analysts* vs. *predictions.* In the abstract stories that experts like to tell, main characters are often abstract nominalizations (boldfaced below):

> 6. Now that we have developed and standardized an index to measure thought disorders, we can quantify how patients respond to different treatments. These **measurements** indicate that **treatments** requiring long-term **hospitalization** do not effectively reduce the number of psychotic episodes among schizophrenic patients.

The nominalizations in that second sentence—*measurements, treatments, hospitalization*—refer to three concepts as familiar to its intended readers as *doctors* and *patients.* Given that audience, the writer would not need to revise them.

In a way, that example undercuts our advice about avoiding nominalizations, because now instead of revising every nominalization, you have to choose which to change into verbs and which to leave alone. For example, the nominalizations in the second sentence of (6) are the same as those in (7a):

> 7a. The **hospitalization** of patients without appropriate **treatment** results in the unreliable **measurement** of outcomes.

But that sentence would profit if those nominalizations were revised into verbs:

> 7b. When we **hospitalize** patients but do not **treat** them appropriately, we cannot **measure** outcomes reliably.

So what we offer here is no iron rule of writing, but rather a principle of diagnosis and revision that you must apply judiciously. In general, however, readers prefer prose whose sentences have subjects that are short, specific, and concrete. And that usually means flesh-and-blood characters.

16.2.5 Abstractions and Characters
The worst problems of abstract prose arise when you create a main character out of a nominalization, use that nominalized

character as the subject of your sentences, but then sprinkle still more nominalizations around it. Here is a passage about two abstract characters, *immediate intention* and *prospective intention*. Those characters are puzzling enough, but note all the other nominalizations in the same passage, complicating that story even more (we underline the subjects and boldface the nominalizations other than "intention"):

> 8a. The **argument** is this: <u>The cognitive component of intention</u> exhibits a high degree of **complexity**. <u>Intention</u> is temporally divisible into two: prospective intention and immediate intention. <u>The cognitive function of prospective intention</u> is the **representation** of a subject's similar past **actions**, his current situation, and his course of future **actions**. That is, <u>the cognitive component of prospective intention</u> is a **plan**. <u>The cognitive function of immediate intention</u> is the **monitoring** and **guidance** of ongoing bodily **movement**. Taken together, <u>these cognitive mechanisms</u> are highly complex. <u>The folk psychological notion of **belief**</u>, however, is an attitude that permits limited **complexity** of content. Thus <u>the cognitive component of intention</u> is something other than folk psychological **belief**.

We can revise this to keep the abstract character *intention,* but if we change unnecessary nominalizations back into verbs and adjectives (they are boldfaced), we create a much clearer passage:

> 8b. <u>My argument</u> is this: <u>The cognitive component of intention</u> is quite **complex**. <u>Intention</u> is temporally divisible into two kinds: prospective intention and immediate intention. <u>The cognitive function of prospective intention</u> **is to represent** how <u>a person</u> has **acted** similarly in the past, his current situation, and how <u>he</u> will **act** in the future. That is, <u>the cognitive component of prospective intention</u> lets him **plan** ahead. <u>The cognitive function of immediate intention</u>, on the other hand, lets him **monitor** and **guide** his body as <u>he</u> **moves** it. Taken together, <u>these cognitive mechanisms</u> are too **complex** to explain in terms of what <u>folk psychology</u> would have us **believe**.

The point: Don't try to change every nominalization into a verb. Some of your central characters may have to be abstractions. But if they are, avoid nominalizations that you do not need. As always, the trick is knowing which ones you need and which you don't (you usually need fewer than you think). Knowing which ones to keep is a skill that comes only from practice and experience.

16.2.6 Picking Main Characters

Having qualified our principle once, we must now complicate it again. If your sentences are readable, your characters will be the subjects of verbs and those verbs will express the crucial actions those characters are involved in. But most stories have several characters, any one of whom we can make more important than others simply by the way we construct sentences. Take our sentence about rain forests:

> 9. If **rain forests** are continuously stripped to serve short-term economic interests, **the entire biosphere** may be damaged.

That sentence tells a story that implies other characters but does not specify them: Who is stripping the forests? More important, does it matter? This story could focus on them, but who are they?

> 9a. If **developers** continue to strip rain forests to serve short-term economic interests, **they** may damage the entire biosphere.
>
> 9b. If **loggers** continue to strip rain forests to serve short-term economic interests, **they** may damage the entire biosphere.
>
> 9c. If **Brazil** continues to strip rain forests to serve short-term economic interests, **it** may damage the entire biosphere.

Which is best? It depends on whom the story *should* be about. As you diagnose sentences, you have two decisions. Whenever possible, put characters in subjects and actions in verbs. But be sure that the character is your *central* character, if only for that sentence.

16.3 A SECOND PRINCIPLE: OLD BEFORE NEW

There is a second principle of reading, diagnosis, and revision even more important than the one we have just explored. Fortunately, the principles are related. Compare the (a) and (b) versions in the following. Which seems easier to get through? Why? (Hint: Look at the way the sentences begin.)

> 10a. Because the naming power of words was distrusted by Locke, he repeated himself often. Seventeenth-century theories of language, especially Wilkins's scheme for a universal language involving the creation of countless symbols for countless meanings, had centered on this naming power. A new era in the study of language that focused on the ambiguous relationship between sense and reference begins with Locke's distrust.

> 10b. Locke often repeated himself because he distrusted the naming power of words. This naming power had been central to seventeenth-century theories of language, especially Wilkins's scheme for a universal language involving the creation of countless symbols for countless meanings. Locke's distrust begins a new era in the study of language, one that focused on the ambiguous relationship between sense and reference.

Most readers prefer (10b). They don't say that (10a) is *too complex* or *inflated*, but that it seems *disjointed*, it does not *flow*—judgmental words that again describe not what is on the page but how readers *feel* about what they are reading.

We can explain what causes those impressions if we again apply the "first six or seven words" test. In the disjointed (a) version, the sentences begin differently from the sentences in the (b) version. The sentences in (10a) begin with information that a reader would find unfamiliar:

> the naming power of words;

> seventeenth-century theories of language;

> a new era in the study of language.

In contrast, the sentences in (10b) begin with information that readers would find familiar:

Locke;

this naming power;

Locke's distrust [a nominalization, but a useful one because it repeats something from the previous sentence].

Some of these are abstractions, but they refer to ideas that appeared in a previous sentence and that readers would recall.

As your readers move from one sentence to the next, they follow your story most easily if they can begin each sentence with a character or idea that is familiar to them, either because you have already mentioned it or because they expect it. From this principle of reading, we can infer our principles of diagnosis and revision:

- Look at the first six or seven words of every sentence.

- Be certain that each opens with information that your readers will find familiar, easy to understand (usually words used before).

- Put close to the ends of your sentences any information that your readers will find new, complex, harder to understand.

This principle cooperates with the one about characters and subjects, because older information usually names a character (after you have introduced it). But should it ever come to a choice between the two, *always choose the principle of old before new.*

16.4 CHOOSING BETWEEN ACTIVE AND PASSIVE

At this point, some of you may recall advice you received from English teachers to avoid passive verbs. That advice is not just misleading; it can be destructive. Rather than worry about active and passive, ask a simpler question: Do your sentences begin with familiar information, preferably a main character? If you put familiar characters in your subjects, you will use the active and passive properly. For example, you may have noticed that one of our earlier examples had passive verbs:

11a. If <u>rain forests</u> continue **to be stripped** to serve short-term economic interests, <u>the entire biosphere</u> may **be damaged**.

Had we followed the standard routine advice, that sentence would have to read:

11b. If <u>loggers</u> continue **to strip** rain forests to serve short-term economic interests, <u>they</u> may **damage** the entire biosphere.

That sentence makes the loggers the main character—fine if the report is about logging and loggers. But if you are telling a story about the gene pool in the Amazon, then the main characters *ought* to be rain forests and the biosphere—and so that sentence *should* be passive.

In English classes, students hear that they should use only active verbs, but in engineering, the natural sciences, and some social sciences, they hear the opposite—use the passive. Most of that advice (based on the alleged interest of scientific objectivity) is equally misleading.

Compare the passive (12a) with the active (12b):

12a. <u>The fluctuations in the current</u> **were measured** at two-second intervals.

12b. <u>We</u> **measured** the fluctuations in the current at two-second intervals.

These sentences are equally objective, but their *stories* differ; one is about fluctuations, the other about the person measuring. The first is supposed to be more "scientific" because it ignores the person and focuses on the current. But the passive in itself is not more objective than the active; it merely implies that the action can be performed by anonymous others who can replicate the procedures. So in this case, the passive is the right choice.

On the other hand, consider this pair of sentences:

13a. <u>It</u> **is suggested** that the fluctuations resulted from the Burnes effect.

13b. <u>We</u> **suggest** that the fluctuations resulted from the Burnes effect.

The active verb in (13b) is not only common in the sciences, but appropriate. The difference? It has to do with the kind of action the verb names. The passive is appropriate when authors refer to actions they perform in the laboratory and that others can replicate: *measure, record, combine,* and so on. But when authors refer to actions that only *they* are entitled to perform—rhetorical actions such as *suggest, prove, claim, argue, show,* and so on—then the authors are the main characters and so they *should* be the subjects of active verbs. Researchers typically use the first person and active verbs at the beginning of journal articles, where they describe how *they* discovered their problem and at the end where they describe how *they* solved it.

16.5 A FINAL PRINCIPLE: COMPLEXITY LAST

We have concentrated on how clauses and sentences begin. Now we'll look at how they end. You can anticipate the principle: If old information goes first, the newest, most complex information goes last. This principle is particularly important in three contexts:

- when you introduce a new technical term;

- when you present a unit of information that is long and complex;

- when you introduce a concept that you intend to develop in what follows.

16.5.1 Introducing Technical Terms

When you introduce a technical term that your readers might be unfamiliar with, construct your sentence so that your technical term appears in the last words. Compare these two:

14a. Calcium blockers can control muscle spasms. Sarcomeres are the small units of muscle fibers in which these drugs work. Two filaments, one thick and one thin, are in each sarcomere. The proteins actin and myosin are contained in the thin filament. When actin and myosin interact, your heart contracts.

14b. Muscle spasms can be controlled with drugs known as calcium blockers. Calcium blockers work in small units of muscle fibers called sarcomeres. Each sarcomere has two filaments, one thick and one thin. The thin filament contains two proteins, actin and myosin. When actin and myosin interact, your heart contracts.

In (14a) all the technical-sounding terms appear early in their sentences; in (14b) the technical terms appear at the end.

16.5.2 Introducing Complex Information

When you express a complex bundle of ideas that you have to state in a long phrase or clause, locate that complexity at the end of its sentence, never at the beginning. Compare these two passages:

15a. There is a second reason historians have concentrated on Darwin rather than Mendel. Hundreds of letters, both personal and scientific, to scores of different recipients, including leading scientific figures, illuminate Darwin's genius. Only ten letters to the botanist Karl Nageli, and a handful to his mother, sister, brother-in-law, and nephew, represent Mendel.

15b. Historians of science have concentrated on Darwin rather than Mendel for a second reason. Darwin's genius is illuminated by hundreds of letters, both personal and scientific, to scores of different recipients, including leading scientific figures. Mendel is represented by only ten letters to the botanist Karl Nageli, and a handful to his mother, sister, brother-in-law, and nephew.

In (15a) the second and third sentences begin with complex units of information, subjects that run on for at least two lines.

In contrast, the subjects in (15b) are short, simple, easy to read—because the passive verbs (*is illuminated* and *is represented*) allowed us to move the short and familiar information to the beginning and the long and complex part to the end. (That's a main purpose of the passive verb.)

In short, if you can recognize when phrases and clauses will seem complex to readers, try to put that complexity not at the beginning of your sentences, but at their end. Unfortunately that's not really easy to do, because you will be too familiar with your own prose to recognize its complexity.

16.5.3 Introducing What Follows

When you are introducing a paragraph, or even a whole section, construct the first sentence of that paragraph so that the key terms of the paragraph are the last words of that sentence. Which of these two sentences would best introduce the excerpt that follows?

> 16a. The political situation changed, because disputes over succession to the throne caused some sort of palace revolt or popular revolution in seven out of eight reigns of the Romanov line after Peter the Great.

> 16b. The political situation changed, because after Peter the Great seven out of eight reigns of the Romanov line were plagued by turmoil over disputed succession to the throne.

> The problems began in 1722, when Peter the Great passed a law of succession that terminated the principle of heredity and required the sovereign to appoint a successor. But because many tsars, including Peter, died before they named successors, those who aspired to rule had no authority by appointment, and so their succession was often disputed by lower-level aristocrats. There was turmoil even when successors were appointed.

Context counts for much here, but most readers who have read these passages feel that (16b) is more cohesive with the rest of the passage. The last few words of (16a) seem relatively unimportant (in a different context, of course, they might be important) and do not introduce the passage that follows as well as (16b).

Therefore, once you've checked the first six or seven words in every sentence, check the last five or six, as well. If those words are not the most important, complex, or weighty, revise so that they are. Look hard at the ends of sentences that introduce paragraphs or even sections.

16.6 SPIT AND POLISH

We've focused on those issues of style especially pertinent to writing research reports, and on principles of diagnosis and revision that help us make prose as readable as possible. There are other principles—sentence length, the right choice of words, concision, and so on. But those are issues pertinent to writing of all kinds, and they are addressed by many books. And, of course, readability is not enough. After you have revised style, structure, and argument, you still have to correct grammar, spelling, punctuation, and citation form. Though important, those matters do not fall within the purview of this book. You can find help in many handbooks.

Our advice about revision may seem very detailed, but if you revise in steps, it is not difficult to follow. The first step is the most important: as you draft, forget about these steps (except for this one). Your first job is to create something to revise. You will never do that if you keep asking yourself whether you should have just used a verb or a noun. If you can't look at every sentence, start with passages where you remember having a hard time explaining your ideas. Whenever you struggle with content, you are likely to tangle up your prose as well. With mature writers, that tangle usually reflects itself in a too complex, "nominalized" style.

For Clarity
Diagnose

1. Highlight the first five or six words in every sentence. Ignore short introductory phrases such as *At first, For the most part,* and so on.

2. Run your eye down the page, looking only at the sequence of highlights to see whether they pick out a consistent set of related words. The words that begin a series of sentences need not be identical, but they should name people or concepts that your readers will see are clearly related. If not, revise.

Revise

1. Identify your main characters, real or conceptual. Make them the subjects of verbs.

2. Look for words ending in *-tion, -ment, -ence,* and so on. If they appear at the beginnings of sentences, turn them into verbs.

For Emphasis
Diagnose

1. Underline the last three or four words in every sentence.

2. In each sentence, identify the words that communicate the newest, most complex, most rhetorically emphatic information; technical-sounding words that you are using for the first time; or concepts that the next several sentences will develop.

Revise

1. Revise your sentences so that those words come last.

PART V

Some Last Considerations

The Ethics of Research

You've heard us do a lot of "preaching" in the last few hundred pages. We've urged you to build a responsible social contract with your readers, to create an ethos that will lead them to trust you, to guard against your inevitable biases in collecting and reporting evidence, to avoid plagiarism, and so on. Now we want to share with you our underlying thinking on such ethical and moral issues, hoping that you will want to think hard about them on your own.

Everything we've said about research reflects our belief that it is a profoundly *social* activity. Reporting research connects us not just to those who will use it, but also to those whose research we used and, through them, to the research that our sources used. And since research is crucial to virtually every facet of our society, that network of social connections among researchers is a defining part of our social fabric. To understand the responsibility we have to those in that network, both to those who rely on our research and to those on whose research we rely, we have to move beyond technique to think about the ethics of civil communication.

Our view of the matter starts with two broad conceptions of the word *ethics:* the range of moral and immoral choices and the construction of bonds within any community. The term *ethical* comes from the Greek *ethos,* meaning either individual *character,* good or bad, or shared *custom* in a community. So far, we have focused on the community-building aspects of research. But more than most social activities, research challenges us to define our individual moral principles and then to make choices that honor or violate them.

At first glance, the academic researcher must seem less tempted to sacrifice principle for gain than, say, a Wall Street researcher evaluating a stock that her firm wants her to push on investors, or a scientist paid by a drug company to "prove" that a product is safe (regardless of whether it is effective). No teacher will pay you to write a report supporting her views or interests, so you probably won't be tempted to fake results to gain fame— like the American researcher who became famous (and powerful) for discovering an HIV virus, when he had in fact "borrowed" it from a lab in France.

Even so, you will face such choices from the very beginning of your project. Some are the obvious *Thou shalt not's:*

• Ethical researchers do not plagiarize or claim credit for the results of others.

• They do not misreport sources or invent results.

• They do not submit data whose accuracy they have reason to question, unless they raise the questions.

• They do not conceal objections that they cannot rebut.

• They do not caricature or distort opposing views.

• They do not destroy or conceal sources and data important for those who follow.

We apply these principles easily enough to obvious cases: the biologist who used india ink to fake "genetic" marks on his mice, the Enron accountants and their auditors at Arthur Andersen who shredded source documents, or the student who submits a paper downloaded from the Internet.

More challenging are those occasions when ethical principles take us beyond any simple moral *Do not* to what we should affirmatively *Do.* When we think about ethical choices in that way, we move beyond simple conflicts between our own self-interest and the honest pursuit of truth, or between what we want and what is good for or at least not harmful to others. If reporting research is genuinely a collaborative effort between readers and writers to find the best solution to shared problems, then the challenge is to find ways to create ethical joining, ethical choices (what we traditionally call *character*) that can help build ethical communities.

Such a challenge raises more questions than we can answer here. Some of those questions we all agree on; others are controversial. The three of us have different answers to some ethical questions. But one thing we might all agree on is that research offers every researcher ethical invitations that, when not just dutifully accepted but *embraced,* can serve the best interests of both researchers and their readers.

- When you try to explain to others why the results of your research *should* change their knowledge, understanding, and beliefs because it is in their best interests to change them, you must examine not only your own understanding and interests, but theirs.

- When you create, however briefly, a community of shared understanding and interest, you set a standard for your work higher than any you would set for yourself alone.

- When you accept the alternative views of your readers, including even their strongest objections and reservations, you help yourself move closer not just to more reliable knowledge, better understanding, and sounder beliefs, but to the dignity and human needs of your readers.

In other words, when you conduct your research and prepare your report as a conversation among equal "characters," all working toward new knowledge and better understanding, the ethical demands you place on yourself should redound to the benefit of all. When you decline the invitation of that conversation, you are likely to harm yourself and possibly those who would depend on your work.

It is this concern for the integrity of the work of the community, combining the narrow moral standards with the larger ethical dimensions, that underscores why researchers condemn plagiarism so strongly. Intentional plagiarism is theft, but of more than words. By not acknowledging a source, the plagiarist steals the recognition that honest researchers should receive, the enhanced respect that a researcher spends a lifetime struggling to earn. And that diminishes the community as a whole, by reducing the value of contributing to the store of knowledge that defines it.

That is true at all levels in all communities, including the undergraduate classroom. The student plagiarist, if successful, steals not only from his sources, but from his colleagues by making other students' work seem worse by comparison. When such intellectual thievery becomes common, the community grows suspicious, then distrustful, then cynical—*So who cares? Everyone does it.* Teachers then have to worry as much about not being tricked as about teaching and learning. What's worse, the plagiarist compromises her own education and so steals from the larger society that has devoted its resources to training her to do reliable work later.

In short, when you report your research to meet the needs of your community, you invite yourself to join that ethical community in a search for the common good. When you respect sources, preserve and acknowledge data that may run against your results, assert claims only as strongly as warranted, acknowledge the limits of your certainty, and meet all the other ethical limits on your report, you move beyond gaining a grade or other material good; you even move beyond simply obeying important moral rules, such as "Never harm anyone by cheating." You earn the larger benefit that comes from bonding with your readers. You discover that research done in the best interests of others is also in your own.

A Postscript for Teachers

In this postscript we want to make explicit what has been implicit throughout. For some, it may seem obvious, but we hope you will join in an effort to improve the national "research scene." Too many teachers, especially teachers of undergraduates, say, *I've given up teaching research.* We hear colleagues tell us that the research papers they get are boring, that students aren't up to the task, that the hard-copy research paper is an outmoded relic of the quaint old days before the Internet, even that no one but ivory-tower academics does research anymore. We think otherwise, of course. We think doing research is the best way to learn to read and think critically. And we are certain that the vast majority of students will have careers in which, if they do not do their own research, they will have to evaluate and depend on the research of others. We know of no way to prepare for that responsibility better than to do research of one's own.

We wrote this book for those who believe—or will consider—two propositions about learning and doing research:

- Students learn to do research well and report it clearly when they take on the perspective of their readers and of the community whose values and practices define competent research and its reporting.

- They learn to manage an important part of that complex mental and social process when they understand how a few key formal features of their reports influence how their readers read and judge them.

These two propositions, we believe, are closely related. By understanding the complementary processes of reading and writing, students plan, perform, and report their research better. Students can use the features that readers expect to guide themselves not only through the process of drafting, but through all the stages of their research. And by understanding what their readers look for in a report, they learn to read the reports of others more critically. The two processes, reading and writing, are mutually supporting.

THE RISKS AND LIMITATIONS OF IMPOSING FORMAL RULES

Emphasizing formal matters, though, carries a risk, especially with beginning researchers. Thoughtless teachers too easily trivialize formal structures into empty drills. Those who teach dancers only to make their feet touch the right marks or pianists only to hit the right keys deprive their students of the sheer pleasure of dancing and playing. Those who teach research as if it were merely learning the right forms for footnotes and bibliography deprive their students of the pleasures of discovery, leading them to join those countless students turned off by Gradgrind formulae, students who might otherwise have blessed the world with their own good research.

If students approach them in the right spirit, the features of an argument are not empty forms to be mindlessly filled, but answers to questions that encourage hard thinking. These patterns help students recognize what is important in the relationship between a researcher, her sources, her disciplinary colleagues, and her readers, a crucial prerequisite to creative and original research.

These patterns, however, will encourage empty imitation if teachers fail to create a rhetorical context that dramatizes for students their social role as researchers, even if at first only in simulation or role-playing. No textbook can fully create that context, because it varies from class to class. It requires a class experience that only imaginative teachers can create.

Only a teacher, understanding his unique students, can orchestrate assignments that create situations whose social dynamic gives point and purpose to research and whose expectations students can recognize and understand. The less experience students have, the more social support teachers must provide before their students can use the formal patterns in productive ways.

ON ASSIGNMENT SCENARIOS: CREATING A GROUND FOR CURIOSITY

Teachers have found many ways of constructing research assignments that give students that necessary support. The most successful have these features:

1. **Good assignments establish outcomes beyond a product to be evaluated.** Good teachers ask students to raise a question or problem that at least they want to resolve, and to support that resolution with reliable and relevant evidence. Effective research assignments then ask students to translate that private interest into a public one, so that they can experience, or at least imagine, a situation in which their readers need the understanding that only they can provide.

The best assignments ask students to write for those who actually need to know or understand something better. Those readers might be a transient community of researchers that a problem creates, as when students do their research for a client outside of class. A senior design class, for example, might address a problem of a local company or civic organization; a music class might write program notes; a history class might investigate the origins of their university or local community.

Less experienced students might write for their classmates, but they might also write for students in another class who could actually use the information that a beginning researcher could provide. They might do preliminary research for those senior design students or for students in a graduate seminar; or they might even write reports back to students still in high school.

Next best are assignments that simulate such situations, in which students assume that other students or a client or even other researchers have a problem that the student researcher will work to resolve. Even in very large classes, students can be organized into small groups who serve as readers with interests that beginning researchers can reasonably address.

2. Good assignments help students learn about their audience. Most students have trouble imagining readers whom they have never met and whose situation they have never experienced. But even when readers are real, students have to imagine their concerns. Biology students with no knowledge or experience of working with a government agency will be unlikely to write a plausible report that meets the concerns of a state EPA administrator. But teachers can help by urging students to imagine that distant audience.

Alternatively, they can turn the class into its own audience by letting students decide what problems need solving, what questions need answering. If students can define the problems they're interested in, they will make the best possible readers for one another's research.

3. Good assignments create scenarios that are rich in contextual information. When students write to resolve the problems of readers known and accessible to them, the assignment creates a scenario with a wealth of reality. Students can investigate, interrogate, and analyze the situation for as long as time and ingenuity allow.

But when it is not practical to locate the project in a real context, the assignment should create as much context as possible. The more information you provide the better. It is seldom possible to anticipate everything students need to know about such a scenario, so it is important to make analysis and discussion of it a part of the writing process. Only when students are working in a social context do they have meaningful choices to make and good reasons to make them. Only then do those choices become rhetorically significant. And only when writers can make rhetorically significant choices will they understand that at the heart of every real writing project is the accurate anticipation of their readers' responses. When students have no choices, either because the project has turned into a mechanical drill or has no rhetorical "scene," doing research and writing it up become merely makework—for you as much as for them.

Again we stress the importance of lively discussion among the students, either in class, if the class is small enough, or in subgroups if the class is large.

4. Good assignments provide interim readers. Few professional researchers call a report finished before they have solicited and evaluated responses, something students need even more. Encourage students to solicit early responses from colleagues, friends, family, even from you. Getting responses is easier if you build opportunities into the assignment itself. Other students can play this role reasonably well, but not if they think that their task is just "editing"—which for them often means rearranging a sentence here and fixing a misspelling there. Have student-responders work through some of the steps in chapters 13–16; you can even create teams of responders, each with responsibility for specific features of the text. Those who provide interim responses must participate in the scenario as imagined readers.

5. As with any real project, good assignments give students time and a schedule of interim deadlines. Research is messy, so it does no good to march students through it lockstep: (1) Select topic, (2) state thesis, (3) write outline, (4) collect bibliography, (5) read and take notes, (6) write report. That caricatures real research. But students need some framework, a schedule of tasks that helps them monitor their progress. They need time for false starts and blind alleys, for revision and reconsideration. They need interim deadlines and stages for sharing and criticizing their progress. Those stages can reflect the various sequences outlined in this book.

RECOGNIZING AND TOLERATING THE INEVITABLE

Students also seriously—sometimes desperately—need other kinds of support, especially recognition of what can be expected of them and tolerance for the predictable missteps of even experienced researchers. Beginners behave in awkward ways, taking suggestions and principles as inflexible rules that they apply mechanically. They work through a topic to a question to the online catalog to a few websites and on to a feeble conclusion, not because they lack imagination or creativity, but because they are struggling to acquire a skill that to them is surpassingly strange. Such awkwardness is an inevitable stage in learning any skill. It passes, but too often after our students have gone on to other classes.

We urge you not to be troubled when a whole class of beginning students produces reports that all look alike. We three have had to learn to be patient with students, as we wait for the delayed gratification that comes when the learners arrive at genuine originality—knowing it will likely come when we are no longer there to see it.

We try to assure students that even if they do not solve their problem, they succeed if they can pose it in a way that convinces us that it is new—at least to them—and arguably *needs* solving. Supporting such a claim often requires more research and more critical ability than merely answering a question. In fact, that kind of proposal paper is often more difficult to write than one in which a student can simply ask a question and answer it.

We know that some students want to use research assignments simply to gather information on a topic, to review a field just to gain control over it. To them, the demand for a significant problem seems artificial. You can only ask them to imagine a reader who is intelligent and possibly interested in their topic but does not have the time to do any research, a reader who is, indeed, in the circumstance they are in.

Finally, different students stand in different relations to the research practices you teach. Advanced students should strive toward the full quality of your own disciplinary practices. But few beginners are yet committed to any research community or to the values that underlie everything in this book. Some will make that commitment early, but most will not. Some never will.

In sum: To teach research well, we have to adapt the steps we've outlined to fit the particular circumstances and needs of the individuals in a class. We hope that students at all levels learn these steps, to identify them in other writing projects and to attempt them on their own. Maybe then they can move toward the kind of research our society so badly needs but too seldom gets.

An Appendix on Finding Sources

There is a vast literature on finding information, only a small part of which we can list. We have divided this list into "General Sources" and "Special Sources," and the "Special Sources" into "Humanities," "Social Sciences," and "Natural Sciences." We then divided each of those areas into their special fields. For each field, we list six kinds of resources:

1. a dictionary that briefly defines concepts and sometimes offers a bibliography;

2. an encyclopedia that gives more extensive overviews and usually a bibliography;

3. a guide to finding resources in the field and using its methodology;

4. bibliographies, abstracts, and indices that list past and current publications in the field;

5. a writing manual for a particular field, if we know of a useful one;

6. a style manual that describes special features of citations.

Some books listed in (3), (5), and (6) may be out of print or available only through interlibrary loan. If there is no date listed for an item, the publication appears annually. Sources available on-line or as a CD-ROM (in addition to or in place of traditional print formats) are so indicated. Online sources for which no URL is given are readily available from multiple online databases.

We said this in the first edition; we say it again here: So rapid is technological change in the information sciences that by the time you read this, new technology will have rendered much of it obsolete. A local bookstore will always have a book to guide you through new technologies. If you do not find what you are looking for on this list, remember that for every field with a name, there is almost certainly a *Dictionary of . . .* , or *Encyclopedia of . . .*

GENERAL SOURCES

1. *Dictionary of American Biography.* New York: Scribner, 1998. Also available as a CD-ROM.

1. Lagassé, Paul, ed. *The Columbia Encyclopedia.* 6th ed. New York: Columbia University Press, 2000. Also available online at http://www.bartleby.com/65.

1. Nicholls, C. S., ed. *The Dictionary of National Biography.* Oxford: Oxford University Press, 1996.

2. Goetz, Phillip W., ed. *The New Encyclopaedia Britannica.* 15th ed. 32 vols. Chicago: Encyclopaedia Britannica, 1987.

3. Balay, Robert, ed. *Guide to Reference Books.* 11th ed. Chicago: American Library Association, 1996.

3. Hacker, Diana, and Barbara Fister. *Research and Documentation in the Electronic Age.* 3rd ed. Boston: Bedford/St. Martin's, 2002.

3. Kane, Eileen, and Mary O'Reilly-de Brún. *Doing Your Own Research: Basic Descriptive Research in the Social Sciences and Humanities.* New York: Marion Boyars Publishing, 2001.

3. Preece, Roy A. *Starting Research: An Introduction to Academic Research and Dissertation Writing.* New York: St. Martin's Press, 1994.

3. Vitale, Philip H. *Basic Tools of Research: An Annotated Guide for Students of English.* 3rd ed., rev. and enl. New York: Barron's Educational Series, 1975.

4. *The Bibliographic Index.* New York: H. W. Wilson. Also available online at http://hwwilsonweb.com.

4. *Biography Reference Bank.* New York: H. W. Wilson. Also available online at http://hwwilsonweb.com and as a CD-ROM.

4. *Books in Print.* New York: R. R. Bowker. Also available online at http://www.booksinprint.com/bip and as a CD-ROM.

4. Brigham, Clarence S. *History and Bibliography of American Newspapers, 1690–1820.* Worcester: American Antiquarian Society, 1947.

4. *Dissertation Abstracts.* New York: SilverPlatter Information. Also available online at http://wwwlib.umi.com/dissertations/search and as a CD-ROM.

4. Gregory, Winifred, ed. *American Newspapers 1821–1936: A Union List of Files Available in the United States and Canada.* New York: H. W. Wilson, 1937.

4. *International Index.* New York: H. W. Wilson.

4. *Lexis-Nexis Academic Universe.* Dayton, Ohio: Available online at http://web.lexis-nexis.com/universe.

4. *Library of Congress Subject Catalog.* Washington, D.C.: Library of Congress. Also available online at http://catalog.loc.gov.

4. *National Newspaper Index.* Menlo Park, Calif.: Information Access. Also available online from multiple sources.

4. *New York Times Index.* New York: New York Times.

4. *Newspapers in Microform.* Washington, D.C.: Library of Congress, 1984.

4. Poole, William Frederick, and William Isaac Fletcher. *Poole's Index to Periodical Literature.* Gloucester: Peter Smith, 1957.

4. *Popular Periodical Index.* Camden, N.J.: Rutgers University.

4. *ProQuest Digital Dissertations.* Ann Arbor: UMI. Also available online at http://wwwlib.umi.com/dissertations/gateway.

4. *Readers' Guide to Periodical Literature.* New York: H. W. Wilson. Also available online at http://hwwilsonweb.com and as a CD-ROM.

4. *Subject Guide to Books in Print.* New York: R. R. Bowker. Also available online at http://www.booksinprint.com/bip and as a CD-ROM.

4. *Wall Street Journal Index.* New York: Dow Jones. Also available online at http://www.il.proquest.com/products/pt-product-WSJ.shtml.

4. *WorldCat.* Dublin, Ohio: Online Computer Library Center. http://www.oclc.org/worldcat.

5. Sternberg, David Joel. *How to Complete and Survive a Doctoral Dissertation.* New York: St. Martin's Griffin, 1981.

5. Strunk, William, and E. B. White. *The Elements of Style.* 4th ed. New York: Longman, 2000.

5. Williams, Joseph M. *Style: Toward Clarity and Grace.* Chicago: University of Chicago Press, 1990.

6. *The Chicago Manual of Style.* 15th ed. Chicago: University of Chicago Press, 2003.

SPECIAL SOURCES
Humanities

1. Murphy, Bruce, ed. *Benet's Reader's Encyclopedia.* 4th ed. New York: HarperCollins, 1996.

3. Kirkham, Sandi. *How to Find Information in the Humanities.* London: Library Association, 1989.

4. Blazek, Ron, and Elizabeth Smith Aversa. *The Humanities: A Selective Guide to Information Sources.* 5th ed. Englewood, Colo.: Libraries Unlimited, 2000. Also available online at http://www.netlibrary.com and as a CD-ROM.

4. *Humanities Index.* New York: H. W. Wilson. Also available online at http://hwwilsonweb.com and as a CD-ROM.

4. Walford, Albert J., Anthony Chalcraft, Ray Prytherch, and Stephen Willis, eds. *Walford's Guide to Reference Material, Volume 3: Generalia, Language and Literature, The Arts.* 7th ed. London: Library Association, 1998.

5. Northey, Margot, and Maurice Legris. *Making Sense in the Humanities: A Student's Guide to Writing and Style.* Toronto: Oxford University Press, 1990.

Art

1. Chilvers, Ian, and Harold Osborne, ed. *The Oxford Dictionary of Art.* 2nd ed. Oxford: Oxford University Press, 2001.

1. Myers, Bernard S., and Trewin Copplestone. *Macmillan Encyclopedia of Art.* Rev. ed. London: Macmillan, 1981.

1. Myers, Bernard S., and Shirley D. Myers, eds. *McGraw-Hill Dictionary of Art.* 5 vols. New York: McGraw-Hill, 1969.

2. *Encyclopedia of World Art.* 17 vols. New York: McGraw-Hill, 1987.

3. Arntzen, Etta, and Robert Rainwater. *Guide to the Literature of Art History.* Chicago: American Library Association, 1980.

3. Jones, Lois Swan. *Art Information: Research Methods and Resources.* 3rd ed. Dubuque: Kendall/Hunt, 1990.

3. Minor, Vernon Hyde. *Art History's History.* 2nd ed. Upper Saddle River, N.J.: Prentice Hall, 2001.

4. *Art Index.* New York: H. W. Wilson. Also available online at http://hwwilsonweb.com and as a CD-ROM.

5. Barnet, Sylvan. *A Short Guide to Writing about Art.* 7th ed. New York: Harper-Collins, 2003.

History

1. Cook, Chris. *A Dictionary of Historical Terms.* 3rd ed. Houndmills, Eng.: Macmillan, 1998.

1. Ritter, Harry. *Dictionary of Concepts in History.* Westport, Conn.: Greenwood Press, 1986.

2. Breisach, Ernst. *Historiography: Ancient, Medieval & Modern.* 2nd ed. Chicago: University of Chicago Press, 1994.

3. Benjamin, Jules R. *A Student's Guide to History.* 8th ed. Boston: Bedford/St. Martin's, 2001.

3. Brundage, Anthony. *Going to the Sources: A Guide to Historical Research and Writing.* 3rd ed. Wheeling: Harlan Davidson, 2002.

3. Frick, Elizabeth. *History: Illustrated Search Strategy and Sources.* 2nd ed. Ann Arbor: Pierian Press, 1995.

3. Kyvig, David E., and Myron A. Marty. *Nearby History: Exploring the Past Around You.* 2nd ed. Walnut Creek, Calif.: AltaMira Press, 2000.

3. Prucha, Francis Paul. *Handbook for Research in American History: A Guide to Bibliographies and Other Reference Works.* 2nd ed. Lincoln: University of Nebraska Press, 1994.

4. *America: History and Life.* Santa Barbara: ABC-CLIO. Also available online at http://serials.abc-clio.com and as a CD-ROM.

4. Blazek, Ron, and Anna H. Perrault. *United States History: A Selective Guide to Information Sources.* Englewood, Colo.: Libraries Unlimited, 1994. Also available online at http://www.netlibrary.com and as a CD-ROM.

4. *Historical Abstracts.* Santa Barbara: ABC-CLIO. Also available online at http://serials.abc-clio.com and as a CD-ROM.

4. Kinnel, Susan K., ed. *Historiography: An Annotated Bibliography of Journal Articles, Books, and Dissertations.* 2 vols. Santa Barbara: ABC-CLIO, 1987.

4. Mott, Frank Luther. *History of American Magazines.* Cambridge: Harvard University Press, 1938.

5. Barzun, Jacques, and Henry F. Graff. *The Modern Researcher.* 5th ed. Boston: Houghton Mifflin, 1992.

6. Marius, Richard, and Melvin E. Page. *A Short Guide to Writing about History.* 4th ed. New York: Longman, 2002.

Literary Studies

1. Abrams, Meyer H. *A Glossary of Literary Terms.* 7th ed. Fort Worth: Harcourt Brace, 1999.

1. Baldick, Chris, ed. *The Concise Oxford Dictionary of Literary Terms.* 2nd ed. Oxford: Oxford University Press, 2001.

1. Brogan, Terry V. F., ed. *The New Princeton Handbook of Poetic Terms.* Princeton: Princeton University Press, 1994.

1. Groden, Michael, and Martin Kreiswirth, eds. *The Johns Hopkins Guide to Literary Theory and Criticism.* Baltimore: Johns Hopkins University Press, 1994.

1. Preminger, Alex, and Terry V. F. Brogan, eds. *The New Princeton Encyclopedia of Poetry and Poetics.* Princeton: Princeton University Press, 1993.

2. Drabble, Margaret, ed. *The Oxford Companion to English Literature.* 6th ed. New York: Oxford University Press, 2000.

2. Hart, James David, and Phillip W. Leininger, eds. *The Oxford Companion to American Literature.* 6th ed. New York: Oxford University Press, 1995.

2. Lentricchia, Frank, and Thomas McLaughlin, eds. *Critical Terms for Literary Study.* 2nd ed. Chicago: University of Chicago Press, 1995.

2. Ward, Sir Adolphus William, A. R. Waller, William Peterfield Trent, John Erskine, Stuart Pratt Sherman, and Carl Van Doren. *The Cambridge History of English and American Literature: An Encyclopedia in Eighteen Volumes.* New York: G. P. Putnam's Sons, 1907–21. Also available online at http://www.bartleby.com/cambridge.

3. Altick, Richard Daniel, and John J. Fenstermaker. *The Art of Literary Research.* 4th ed. New York: Norton, 1993.

4. *Abstracts of English Studies.* Boulder: National Council of Teachers of English.

4. Blanck, Jacob. *Bibliography of American Literature.* New Haven: Yale University Press, 1955. Also available online at http://lion.chadwyck.co.uk and as a CD-ROM.

4. *MLA International Bibliography of Books and Articles on the Modern Languages and Literature.* New York: MLA. Also available online from multiple sources and as a CD-ROM.

5. Barnet, Sylvan, and William E. Cain. *A Short Guide to Writing about Literature*. 9th ed. New York: Longman, 2002.

5. Griffith, Kelley. *Writing Essays about Literature: A Guide and Style Sheet*. 6th ed. Fort Worth: Harcourt College Publishers, 2002.

6. Gibaldi, Joseph. *MLA Handbook for Writers of Research Papers*. 5th ed. New York: MLA, 1999.

Music

1. Randel, Don Michael, ed. *The New Harvard Dictionary of Music*. Cambridge: Belknap Press of Harvard University Press, 1986.

1. Sadie, Stanley, and John Tyrrell, eds. *The New Grove Dictionary of Music and Musicians*. 2nd ed. 29 vols. New York: Macmillan, 2001. Also available online at http://www.grovemusic.com/index.html.

2. Sadie, Stanley, ed. *The Norton/Grove Concise Encyclopedia of Music*. Rev. and enl. ed. New York: W. W. Norton, 1994.

3. Brockman, William S. *Music: A Guide to the Reference Literature*. Littleton, Colo.: Libraries Unlimited, 1987.

3. Duckles, Vincent H., Ida Reed, and Michael A. Keller, eds. *Music Reference and Research Materials: An Annotated Bibliography*. 5th ed. New York: Schirmer Books, 1997.

4. *The Music Index*. Detroit: Information Service. Also available online at http://www.hppmusicindex.com/home.asp and as a CD-ROM.

4. *RILM Abstracts of Music Literature*. New York: RILM International Center. Also available online from multiple sources and as a CD-ROM.

5. Druesedow, John E., Jr. *Library Research Guide to Music: Illustrated Search Strategy and Sources*. Ann Arbor: Pierian Press, 1982.

5. Wingell, Richard. *Writing about Music: An Introductory Guide*. 3rd ed. Upper Saddle River, N.J.: Prentice Hall, 2002.

6. Holoman, D. Kern. *Writing about Music: A Style Sheet from the Editors of* 19th Century Music. Berkeley: University of California Press, 1988.

Philosophy

1. Blackburn, Simon. *The Oxford Dictionary of Philosophy*. New York: Oxford University Press, 1996. Also available online at http://www.oxfordreference.com/views/BOOK_SEARCH.html?book=t98&subject=s22.

1. Urmson, J. O., and Jonathan Reé, eds. *Concise Encyclopedia of Western Philosophy and Philosophers*. New ed., completely rev. London: Routledge, 1991.

2. Edwards, Paul. *The Encyclopedia of Philosophy*. 8 vols. New York: Simon & Schuster Macmillan, 1996.

2. Parkinson, George H. R. *The Handbook of Western Philosophy*. New York: Macmillan, 1988.

3. List, Charles J., and Stephen H. Plum. *Library Research Guide to Philosophy*. Ann Arbor: Pierian Press, 1990.

4. *The Philosopher's Index*. Bowling Green, Ky.: Bowling Green University

Press. Also available online at http://www.silverplatter.com/catalog.phil.htm and as a CD-ROM.

5. Martinich, Aloysius. *Philosophical Writing: An Introduction*. 2nd ed. Cambridge: Blackwell, 1996.

5. Watson, Richard A. *Writing Philosophy: A Guide to Professional Writing and Publishing*. Carbondale: Southern Illinois University Press, 1992.

Women's Studies

1. Bataille, Gretchen M., and Laurie Lisa, eds. *Native American Women: A Biographical Dictionary*. 2nd ed. New York: Routledge, 2001.

1. Mills, Jane. *Womanwords: A Dictionary of Words about Women*. New York: Henry Holt, 1993.

1. Salem, Dorothy C., ed. *African American Women: A Biographical Dictionary*. New York: Garland, 1993.

1. Uglow, Jennifer S., and Frances Hinton, eds. *The Continuum Dictionary of Women's Biography*. New enl. ed. New York: Continuum, 1989.

2. Hine, Darlene Clark, Elsa Barkley Brown, and Rosalyn Terborg-Penn, eds. *Black Women in America: An Historical Encyclopedia*. 2 vols. Bloomington: Indiana University Press, 1994.

2. Tierney, Helen, ed. *Women's Studies Encyclopedia*. Rev. and enl. ed. 3 vols. Westport, Conn.: Greenwood Press, 1999. Also available online at http://www.gem.greenwood.com and as a CD-ROM.

2. Willard, Frances E., and Mary A. Livermore, eds. *American Women: Fifteen Hundred Biographies with Over 1,400 Portraits*. 2 vols. Rev. ed. Detroit: Gale Research, 1973.

3. Searing, Susan E. *Introduction to Library Research in Women's Studies*. Boulder: Westview Press, 1985.

4. *Women's Studies Abstracts*. Rush, N.Y.: Rush Publishing.

Social Sciences

1. *Statistical Abstract of the United States*. Washington, D.C.: Bureau of the Census. Also available online at http://www.census.gov/statab/www and as a CD-ROM.

2. Sills, David, ed. *International Encyclopedia of the Social Sciences*. 19 vols. New York: Macmillan, 1991.

3. Light, Richard J., and David B. Pillemer. *Summing Up: The Science of Reviewing Research*. Cambridge: Harvard University Press, 1984.

3. Øyen, Else, ed. *Comparative Methodology: Theory and Practice in International Social Research*. Newbury Park, Calif.: Sage, 1990.

4. Day, Alan, and Michael Walsh, eds. *Walford's Guide to Reference Material, Volume 2: Social and Historical Sciences, Philosophy and Religion*. 8th ed. London: Library Association, 2000.

4. *Social Sciences Index*. New York: H. W. Wilson. Also available online at http://hwwilsonweb.com and as a CD-ROM.

5. Becker, Howard S. *Writing for Social Scientists: How to Start and Finish Your Thesis, Book, or Article.* Chicago: University of Chicago Press, 1986.

5. Bell, Judith. *Doing Your Research Project: A Guide for First-time Researchers in Education and Social Science.* 3rd ed. Philadelphia: Open University Press, 1999.

5. Krathwohl, David R. *How to Prepare a Research Proposal: Guidelines for Funding and Dissertations in the Social and Behavioral Sciences.* 3rd ed. Syracuse: Syracuse University Press, 1988.

5. Northey, Margot, Lorne Tepperman, and James Russell. *Making Sense in the Social Sciences: A Student's Guide to Research, Writing and Style.* 2nd ed. Toronto: Oxford University Press, 2001.

Anthropology

1. Winthrop, Robert H. *Dictionary of Concepts in Cultural Anthropology.* New York: Greenwood Press, 1991.

2. Ingold, Tim, ed. *Companion Encyclopedia of Anthropology.* New York: Routledge, 1994.

2. Levinson, David, ed. *Encyclopedia of World Cultures.* Boston: G. K. Hall, 1996.

2. Levinson, David, and Melvin Ember, eds. *Encyclopedia of Cultural Anthropology.* New York: Henry Holt, 1996.

3. Bernard, H. Russell. *Handbook of Methods in Cultural Anthropology.* Walnut Creek, Calif.: AltaMira Press, 1998.

3. Bernard, H. Russell. *Research Methods in Anthropology: Qualitative and Quantitative Methods.* 3rd ed. Walnut Creek, Calif.: AltaMira Press, 2002.

3. *Current Topics in Anthropology: Theory, Methods, and Content.* 8 vols. Reading, Mass.: Addison-Wesley.

3. Glenn, James R. *Guide to the National Anthropological Archives, Smithsonian Institution.* Rev. and enl. Washington, D.C.: National Anthropological Archives, 1996.

4. *Abstracts in Anthropology.* Amityville, N.Y.: Baywood Publishing [formerly, Westport, Conn.: Greenwood Press].

4. *Annual Review of Anthropology.* Palo Alto, Calif.: Annual Reviews. Also available online at http://anthro.AnnualReviews.org/contents-by-date.o.shtml.

5. Poggie, John J., Jr., Billie R. DeWalt, and William W. Dressler, eds. *Anthropological Research: Process and Application.* Albany: State University of New York Press, 1992. Also available online at http://www.netlibrary.com.

Business

1. Link, Albert N. *Link's International Dictionary of Business Economics.* Chicago: Probus, 1993.

1. Nisberg, Jay N. *The Random House Dictionary of Business Terms.* New York: Random House, 1992.

1. Wiechmann, Jack G., and Laurence Urdang, eds. *NTC's Dictionary of Advertising.* 2nd ed. Lincolnwood, Ill.: NTC, 1992.

2. *Encyclopedia of American Business History and Biography.* New York: Facts on File, 1995.

2. *The Lifestyle Market Analyst: A Reference Guide for Consumer Market Analysis.* Wilmette, Ill.: SRDS.

3. Cousins, Jill, and Lesley Robinson, eds. *The Online Manual: A Practical Guide to Business Databases.* 2nd ed. New York: Blackwell, 1993.

3. Daniells, Lorna M. *Business Information Sources.* 3rd ed. Berkeley: University of California Press, 1993.

3. Kervin, John B. *Methods for Business Research.* New York: HarperCollins, 1992.

3. Sekaran, Uma. *Research Methods for Business: A Skill-Building Approach.* 4th ed. New York: Wiley, 2002.

3. Woy, James B., ed. *Encyclopedia of Business Information Sources.* 17th ed. Detroit: Gale Research, 2002.

4. *Business Periodicals Index.* New York: H. W. Wilson. Also available online at http://hwwilsonweb.com and as a CD-ROM.

5. Farrell, Thomas J., and Charlotte Donabedian. *Writing the Business Research Paper: A Complete Guide.* Durham: Carolina Academic Press, 1991.

6. Vetter, William. *Business Law, Legal Research, and Writing: Handbook.* Needham Heights, Mass.: Ginn Press, 1991.

Communications and Journalism

1. Newton, Harry. *Newton's Telecom Dictionary.* 18th ed. New York: CMP Books, 2002.

1. Weik, Martin H. *Communications Standard Dictionary.* 3rd ed. New York: Chapman & Hall, 1996. Also available as a CD-ROM.

1. Weiner, Richard. *Webster's New World Dictionary of Media and Communications.* Rev. ed. New York: Macmillan, 1996.

2. Barnouw, Erik, ed. *International Encyclopedia of Communications.* 4 vols. New York: Oxford University Press, 1989.

2. Paneth, Donald. *The Encyclopedia of American Journalism.* New York: Facts on File, 1983.

2. Stern, Jane, and Michael Stern. *Encyclopedia of Pop Culture: An A to Z Guide of Who's Who and What's What, from Aerobics and Bubble Gum to Valley of the Dolls and Moon Unit Zappa.* New York: Harper Perennial, 1992.

3. Block, Eleanor S., and James K. Bracken. *Communication and the Mass Media: A Guide to the Reference Literature.* Englewood, Colo.: Libraries Unlimited, 1991.

3. Blum, Eleanor, and Frances Goins Wilhoit. *Mass Media Bibliography: An Annotated Guide to Books and Journals for Research and Reference.* 3rd ed. Urbana: University of Illinois Press, 1990.

3. Cates, Jo A. *Journalism: A Guide to the Reference Literature.* 2nd ed. Englewood, Colo.: Libraries Unlimited, 1997. Also available online at http://www.netlibrary.com.

3. Sterling, Christopher H., James K. Bracken, and Susan M. Hill, eds. *Mass Communications Research Resources.* Mahwah, N.J.: Erlbaum, 1998.

4. *Communications Abstracts.* Los Angeles: University of California Press.

4. Matlon, Ronald J., and Sylvia P. Ortiz, eds. *Index to Journals in Communication Studies through 1995.* Annadale, Va.: National Communication Association, 1997.

6. Goldstein, Norm, ed. *The Associated Press Stylebook and Libel Manual: With Appendixes on Photo Captions, Filing the Wire.* 34th ed. New York: Associated Press, 1999.

Economics

1. Pearce, David W., ed. *MIT Dictionary of Modern Economics.* 4th ed. Cambridge: MIT Press, 1992.

2. Eatwell, John, Murray Milgate, Peter K. Newman, and Sir Robert Harry Inglis Palgrave, eds. *The New Palgrave: A Dictionary of Economics.* 4 vols. New York: Palgrave, 2002.

2. Greenwald, Douglas, ed. *The McGraw-Hill Encyclopedia of Economics.* 2nd ed. New York: McGraw-Hill, 1994.

3. Fletcher, John, ed. *Information Sources in Economics.* 2nd ed. London: Butterworths, 1984.

3. Johnson, Glenn L. *Research Methodology for Economists: Philosophy and Practice.* New York: Macmillan, 1986.

4. *Journal of Economic Literature.* Nashville: American Economic Association. Also available online at http://www.e-jel.org and as a CD-ROM.

5. McCloskey, Deirdre [Donald] N. *The Writing of Economics.* New York: Macmillan, 1987.

Education

1. Barrow, Robin, and Geoffrey Milburn. *A Critical Dictionary of Educational Concepts: An Appraisal of Selected Ideas and Issues in Educational Theory and Practice.* 2nd ed. New York: Teacher's College Press, 1990.

1. Lawton, Denis, and Peter Gordon. *Dictionary of Education.* 2nd ed. London: Hodder & Stoughton, 1996.

2. Alkin, Marvin C., ed. *Encyclopedia of Educational Research.* 6th ed. 4 vols. New York: Macmillan, 1992.

2. Husen, Torsten, and T. Neville Postlethwaite, eds. *International Encyclopedia of Education.* 2nd ed. 12 vols. New York: Pergamon, 1994.

3. Bausell, R. Barker. *Advanced Research Methodology: An Annotated Guide to Sources.* Metuchen, N.J.: Scarecrow Press, 1991.

3. Keeves, John P., ed. *Educational Research, Methodology, and Measurement: An International Handbook.* 2nd ed. New York: Pergamon, 1997.

3. O'Brien, Nancy P. *Education: A Guide to Reference and Information Sources.* 2nd ed. Englewood, Colo.: Libraries Unlimited, 2000. Also available online at http://www.netlibrary.com.

4. *Education Index.* New York: H. W. Wilson. Also available online at http://hwwilsonweb.com and as a CD-ROM.

4. *The ERIC Database.* Lanham, Md.: Educational Resources Information Center. Also available online at http://www.eric.ed.gov and as a CD-ROM.

5. Tuckman, Bruce W. *Conducting Educational Research.* 5th ed. Fort Worth: Harcourt Brace, 1999.

6. Carver, Ronald P. *Writing a Publishable Research Report: In Education, Psychology, and Related Disciplines.* Springfield, Ill.: C. C. Thomas, 1984.

Geography

1. Witherick, M. E., Simon Ross, and John Small. *A Modern Dictionary of Geography.* 4th ed. London: Arnold, 2001.

2. Dunbar, Gary S. *Modern Geography: An Encyclopedic Survey.* New York: Garland, 1991.

2. Parker, Sybil P., ed. *World Geographical Encyclopedia.* 5 vol. New York: McGraw-Hill, 1995.

3. Walford, Nigel. *Geographical Data Analysis.* New York: Wiley, 1995.

4. Conzen, Michael P., Thomas A. Rumney, and Graeme Wynn. *A Scholar's Guide to Geographical Writing on the American and Canadian Past.* Chicago: University of Chicago Press, 1993.

4. *Current Geographical Publications.* New York: American Geographical Society of New York.

4. *Geographical Abstracts.* Norwich, Eng.: Geo Abstracts Ltd.

4. Okuno, Takashi. *A World Bibliography of Geographical Bibliographies.* Ibaraki, Japan: Tsukuba Institute of Geoscience, 1992.

5. Durrenberger, Robert W. *Geographical Research and Writing.* New York: Crowell, 1971.

6. Northey, Margot, and David B. Knight. *Making Sense in Geography and Environmental Studies: A Student's Guide to Research, Writing, and Style.* 2nd ed. Toronto: Oxford University Press, 2000.

Law

1. Curzon, L. B. *Dictionary of Law.* 6th ed. Harlow: Pearson Education, 2002.

1. Garner, Bryan A., ed. *Black's Law Dictionary.* 7th ed. St. Paul, Minn.: West Publishing, 1999.

2. Baker, Brian L., and Patrick J. Petit, eds. *Encyclopedia of Legal Information Sources.* 2nd ed. Detroit: Gale Research, 1993.

2. *Corpus Juris Secundum.* St. Paul, Minn.: West Publishing.

2. *West's Encyclopedia of American Law.* 12 vols. St. Paul, Minn.: West Publishing, 1998.

3. Campbell, Enid Mona, Lee Poh-York, and Joycey G. Tooher. *Legal Research: Materials and Methods.* 4th ed. North Ryde, Aus.: LBC Information Services, 1996.

4. *Current Index to Legal Periodicals.* Seattle: M. G. Gallagher Law Library and Washington Law Review. Also available online at http://lib.law.washington.edu/cilp/cilp.html.

4. *Encyclopedia of Legal Information Sources: A Bibliographic Guide.* 2nd ed. Detroit: Gale Research, 1993.

4. *Index to Legal Periodicals & Books.* New York: H. W. Wilson. Also available online at http://hwwilsonweb.com and as a CD-ROM.

4. *Lawdesk.* Rochester: Lawyers Cooperative Publications. CD-ROM.

5. Bast, Carol M. *Legal Research and Writing.* Albany: Delmar Publishers, 1995.

6. *The Bluebook: A Uniform System of Citation.* 17th ed. Cambridge: Harvard Law Review Association, 2000.

Political Science

1. Robertson, David. *The Penguin Dictionary of Politics.* 2nd ed. London: Penguin, 1993.

2. *The Almanac of American Politics.* Washington, D.C.: National Journal. Also available online at http://nationaljournal.com/members/almanac.

2. Hawkesworth, Mary E., and Maurice Kogan, eds. *Encyclopedia of Government and Politics.* 2 vols. New York: Routledge, 1992.

2. Lal, Shiv, ed. *International Encyclopedia of Politics and Laws.* 17 vols. New Delhi: Election Archives, 1987.

2. Miller, David, ed. *The Blackwell Encyclopaedia of Political Thought.* New York: Blackwell, 1987. Also available online at http://www.netlibrary.com.

3. Holler, Frederick L., ed. *Information Sources of Political Science.* 4th ed. Santa Barbara: ABC-CLIO, 1986.

3. Johnson, Janet Buttolph, Richard A. Joslyn, and H. T. Reynolds. *Political Science Research Methods.* 4th ed. Washington, D.C.: Congressional Quarterly Press, 2001.

4. *ABC: Pol Sci.* Santa Barbara: ABC-CLIO. Also available as a CD-ROM.

4. *PAIS International Journals Indexed.* New York: Public Affairs Information Service. Also available online from multiple sources and as a CD-ROM.

5. Biddle, Arthur W., Kenneth M. Holland, and Toby Fulwiler. *Writer's Guide: Political Science.* Lexington, Mass.: D. C. Heath, 1987.

5. Lovell, David W., and Rhonda Moore. *Essay Writing and Style Guide for Politics and the Social Sciences.* Sydney: Australasian Political Studies Association, 1992.

Psychology

1. Eysenck, Michael, ed. *The Blackwell Dictionary of Cognitive Psychology.* Cambridge: Blackwell, 1994.

1. Stratton, Peter, and Nicky Hayes. *A Student's Dictionary of Psychology.* 3rd ed. London: Arnold, 1999.

1. Wolman, Benjamin B., ed. *Dictionary of Behavioral Science.* 2nd ed. San Diego: Academic Press, 1989.

2. Colman, Andrew M., ed. *Companion Encyclopedia of Psychology.* 2 vols. New York: Routledge, 1994.

2. Corsini, Raymond J., ed. *Encyclopedia of Psychology.* 3rd ed. 4 vols. New York: Wiley, 2001.

3. Breakwell, Glynis M., Sean Hammond, and Chris Fife-Schaw. *Research Methods in Psychology.* 2nd ed. Thousand Oaks, Calif.: Sage, 2000.

3. Elmes, David G., Barry H. Kantowitz, and Henry L. Roediger III. *Research Methods in Psychology.* 7th ed. Belmont, Calif.: Wadsworth, 2002.

3. Reed, Jeffrey G., and Pam M. Baxter. *Library Use: A Handbook for Psychology.* 2nd ed. Washington, D.C.: American Psychological Association, 1992.

3. Wilson, Christopher. *Research Methods in Psychology: An Introductory Laboratory Manual.* Dubuque, Iowa: Kendall-Hunt, 1990.

4. *Annual Review of Psychology.* Palo Alto, Calif.: Annual Reviews.

4. *Compact Cambridge MEDLINE.* Bethesda, Md.: NLM by Cambridge Scientific Abstracts. Also available online at http://www.ncbi.nlm.nih.gov/entrez/query.fcgi.

4. *NASPSPA Abstracts.* Champaign, Ill.: Human Kinetics Publishers.

4. *Psychological Abstracts.* Lancaster, Pa.: American Psychological Association. Also available online at http://www.apa.org/psycinfo and as a CD-ROM with the title *The PsycLit Database.*

4. *The Web of Science Citation Databases.* Philadelphia: Institute for Scientific Information. Also available online at http://isio.isiknowledge.com.

5. Solomon, Paul R. *A Student's Guide to Research Report Writing in Psychology.* Glenview, Ill.: Scott Foresman, 1985.

5. Sternberg, R. J. *The Psychologist's Companion: A Guide to Scientific Writing for Students and Researchers.* 3rd ed. New York: Cambridge University Press, 1993.

6. *Publication Manual of the American Psychological Association.* 5th ed. Washington, D.C.: American Psychological Association, 2001.

Religion

1. Pye, Michael, ed. *Continuum Dictionary of Religion.* New York: Continuum, 1994.

2. Eliade, Mircea, ed. *Encyclopedia of Religion.* 16 vols. New York: Macmillan, 1995.

3. Kennedy, J. *Library Research Guide to Religion and Theology: Illustrated Search Strategy and Sources.* 2nd ed., rev. Ann Arbor: Pierian, 1984.

4. Brown, David, and Richard Swinbourne. *A Selective Bibliography of the Philosophy of Religion.* Oxford: Sub-Faculty of Philosophy, 1995.

4. Chinvamu, Salms. *An Annotated Bibliography on Religion.* Malawi, Africa: Malawi Library Association, 1993.

4. O'Brien, Betty A., and Elmer J. O'Brien, eds. *Religion Index Two: Festschriften, 1960–1969.* Chicago: American Theological Library Association, 1980. Also available online at http://purl.org/atlaonline and as a CD-ROM.

4. *Religion Index One: Periodicals.* Chicago: American Theological Library Association. Also available online at http://purl.org/atlaonline and as a CD-ROM.

4. *Religion Index Two: Multi-author Works.* Chicago: American Theological

Library Association. Also available online at http://purl.org/atlaonline and as a CD-ROM.

Sociology

1. Abercrombie, Nicholas, Stephen Hill, and Bryan S. Turner. *The Penguin Dictionary of Sociology.* 4th ed. London: Penguin, 2000. 1. Gordon, Marshall, ed. *The Concise Oxford Dictionary of Sociology.* New York: Oxford University Press, 1994.

2. Borgatta, Edgar F., ed. *Encyclopedia of Sociology.* 2nd ed. 5 vols. New York: Macmillan, 2000.

2. Smelser, N., ed. *Handbook of Sociology.* Newbury Park, Calif.: Sage, 1988.

3. Aby, Stephen H., ed. *Sociology: A Guide to Reference and Information Sources.* 2nd ed. Englewood, Colo.: Libraries Unlimited, 1997.

3. Lieberson, Stanley. *Making It Count: The Improvement of Social Research and Theory.* Berkeley: University of California Press, 1985.

4. *Annual Review of Sociology.* Palo Alto, Calif.: Annual Reviews.

4. *ASSIA.* London: Bowker Saur. Also available online at http://www.assianet.co.uk/csa/factsheets/assia.shtml and as a CD-ROM.

4. *Social Sciences Index.* New York: H. W. Wilson. Also available online at http://hwwilsonweb.com and as a CD-ROM.

4. *Social Science Research.* Philadelphia: Institute for Scientific Information.

4. *Sociological Abstracts.* New York: Sociological Abstracts. Also available online from multiple sources and as a CD-ROM.

5. Sociology Writing Group. *A Guide to Writing Sociology Papers.* 5th ed. New York: Worth, 2001.

5. Tomovic, Vladislav A., ed. *Definitions in Sociology: Convergence, Conflict, and Alternative Vocabularies: A Manual for Writers of Term Papers, Research Reports, and Theses.* St. Catharines, Ont.: Diliton Publications, 1979.

Natural Sciences

1. *McGraw-Hill Science and Technical Reference Set.* New York: McGraw-Hill, 1992. CD-ROM, version 2.0.

1. Morris, Christopher, ed. *Academic Press Dictionary of Science and Technology.* San Diego: Academic Press, 1992.

1. Walker, Peter M. B., ed. *Chambers Dictionary of Science and Technology.* Edinburgh, U.K.: Chambers, 1999.

2. *McGraw-Hill Dictionary of Scientific and Technical Terms.* 6th ed. New York: McGraw-Hill, 2002.

2. *McGraw-Hill Multimedia Encyclopedia of Science and Technology.* New York: McGraw-Hill. CD-ROM, version 2.0.

2. *Van Nostrand's Scientific Encyclopedia.* 9th ed. New York: Wiley, 2002.

3. *Directory of Technical and Scientific Directories: A World Bibliographic Guide to Medical, Agricultural, Industrial, and Natural Science Directories.* 6th ed. Phoenix: Oryx Press, 1989.

3. Hurt, Charlie Deuel. *Information Sources in Science and Technology*. 3rd ed. Englewood, Colo.: Libraries Unlimited, 1998. Also available online at http://www.netlibrary.com.

3. Nielsen, Harry A. *Methods of Natural Science: An Introduction*. Englewood Cliffs, N.J.: Prentice Hall, 1967.

4. *Applied Science & Technology Index*. New York: H. W. Wilson. Also available online at http://hwwilsonweb.com and as a CD-ROM.

4. *General Science Index*. New York: H. W. Wilson. Also available online at http://hwwilsonweb.com and as a CD-ROM.

4. *ISI Web of Science: Science Citation Index*. Philadelphia: Institute for Scientific Information. Also available online at http://isio.isiknowledge.com.

5. Booth, Vernon. *Communicating in Science: Writing a Scientific Paper and Speaking at Scientific Meetings*. 2nd ed. New York: Cambridge University Press, 1993.

5. Gilpin, Andrea A., and Patricia Patchet-Golubev. *A Guide to Writing in the Sciences*. Toronto: University of Toronto Press, 2000.

5. Valiela, Ivan. *Doing Science: Design, Analysis, and Communication of Scientific Research*. New York: Oxford University Press, 2001.

5. Wilson, Anthony, et al. *Handbook of Science Communication*. Bristol: Institute of Physics Publishing, 1998.

6. Rubens, Phillip, ed. *Science and Technical Writing: A Manual of Style*. 2nd ed. New York: Routledge, 2001.

6. Style Manual Comm., Council of Biology Editors. *Scientific Style and Format: The CBE Manual for Authors, Editors, and Publishers*. 6th ed. Cambridge: Cambridge University Press, 1994.

Biology

1. Allaby, Michael, ed. *The Oxford Dictionary of Natural History*. New York: Oxford University Press, 1985.

1. Lawrence, Eleanor, ed. *Henderson's Dictionary of Biological Terms*. 12th ed. New York: Prentice Hall, 2000.

1. *Oxford Dictionary of Biology*. 4th ed. Oxford: Oxford University Press, 2000.

1. Singleton, Paul, and Diana Sainsbury. *Dictionary of Microbiology and Molecular Biology*. 3rd ed. New York: Wiley, 2001.

2. *Biology Encyclopedia*. New York: HarperCollins, 1991. Videodisc.

2. Creighton, Thomas E., ed. *Encyclopedia of Molecular Biology*. 4 vols. New York: Wiley, 1999.

2. Dulbecco, Renato, ed. *Encyclopedia of Human Biology*. 9 vols. 2nd ed. San Diego: Academic Press, 1997.

3. Roper, Fred W., and Jo Anne Boorkman. *Introduction to Reference Sources in the Health Sciences*. 3rd ed. Metuchen, N.J.: Scarecrow, 1994.

3. Wyatt, H. V., ed. *Information Sources in the Life Sciences*. 4th ed. London: Bowker-Saur, 1997.

4. *Biological Abstracts*. Philadelphia: Biological Abstracts. Also available online from multiple sources and as a CD-ROM.

4. *Biological and Agricultural Index.* New York: H. W. Wilson. Also available online at http://hwwilsonweb.com and as a CD-ROM.

4. *Environment Abstracts.* Bethesda, Md.: Congressional Information Service. Also available online from multiple sources and as a CD-ROM.

4. *ISI Web of Science: Science Citation Index.* Philadelphia: Institute for Scientific Information. Also available online at http://isio.isiknowledge.com.

5. McMillan, Victoria E. *Writing Papers in the Biological Sciences.* 3rd ed. Boston: Bedford/St. Martin's, 2001.

6. *CBE Style Manual: A Guide for Authors, Editors, and Publishers in the Biological Sciences.* 5th ed. Bethesda, Md.: Council of Biology Editors, 1983.

Chemistry

1. *CRC Handbook of Chemistry and Physics.* 76th ed. Boca Raton: CRC Press, 1995.

1. *Hawley's Condensed Chemical Dictionary.* Rev. by Richard J. Lewis Sr. 14th ed. New York: Wiley, 2001.

2. Kroschwitz, Jacqueline, and Mary Howe-Grant, eds. *Kirk-Othmer Encyclopedia of Chemical Technology.* 4th ed. New York: Wiley, 1991.

2. Meyers, Robert A., ed. *Encyclopedia of Physical Science and Technology.* 3rd ed. San Diego: Academic Press, 2002.

3. Leslie, Davies. *Efficiency in Research, Development, and Production: The Statistical Design and Analysis of Chemical Experiments.* Cambridge: Royal Society of Chemistry, 1993.

3. Wiggins, Gary. *Chemical Information Sources.* New York: McGraw-Hill, 1991.

4. *ACS Publications.* Washington, D.C.: American Chemical Society.

4. *Chemical Abstracts.* Washington, D.C.: American Chemical Society. Also available online at http://www.cas.org/stn.html and as a CD-ROM.

4. *Composite Index for CRC Handbooks.* 3rd ed. Boca Raton: CRC Press, 1991.

4. *CRC Handbook of Chemistry and Physics.* Cleveland: CRC Press.

4. *CrossFire Beilstein.* Frankfurt: MDL Information Systems. Also available online at http://www.beilstein.com/products/xfire.

4. *ISI Web of Science: Science Citation Index.* Philadelphia: Institute for Scientific Information. Also available online at http://isio.isiknowledge.com.

4. *ScienceDirect.* New York: Elsevier. Also available online at http://www.sciencedirect.com.

5. Dodd, Janet S., ed. *The ACS Style Guide: A Manual for Authors and Editors.* 2nd ed. Washington, D.C.: American Chemical Society, 1997.

6. Schoenfeld, Robert. *The Chemist's English, with "Say It in English, Please!"* 3rd rev. ed. New York: VCH, 1989.

Computer Sciences

1. LaPlante, Phillip A. *Dictionary of Computer Science, Engineering, and Technology.* Boca Raton: CRC Press, 2001.

1. South, David W. *The Computer and Information Science and Technology Abbreviations and Acronyms Dictionary.* Boca Raton: CRC Press, 1994.
1. Spencer, Donald. *Webster's New World Dictionary of Computer Terms.* 5th ed. New York: Macmillan, 1994.
2. Henderson, Harry. *Encyclopedia of Computer Science and Technology.* New York: Facts on File, 2003.
2. Ralston, Anthony, and Edwin D. Reilly. *Encyclopedia of Computer Science.* 4th ed. New York: International Thomson Computer Press, 2000.
3. Ardis, Susan B. *A Guide to the Literature of Electrical and Electronics Engineering.* Littleton, Colo.: Libraries Unlimited, 1987.
4. *Applied Science and Technology Index.* New York: H. W. Wilson. Also available online at http://hwwilsonweb.com and as a CD-ROM.
4. *Directory of Library Automation Software, Systems, and Services.* Medford, N.J.: Learned Information.
5. Eckstein, C. J. *Style Manual for Use in Computer-Based Instruction.* Brooks Air Force Base, Tex.: Air Force Human Resources Laboratory, Air Force Systems Command, 1990.

Geology and Earth Sciences

1. Clark, John O. E., and Stella Stiegeler, eds. *The Facts on File Dictionary of Earth Science.* New York: Facts on File, 2000.
1. Jackson, Julia A., ed. *Glossary of Geology.* 4th ed. Alexandria, Va.: American Geological Institute, 1997.
2. Bishop, Arthur C., Alan R. Woolley, and William R. Hamilton. *Cambridge Guide to Minerals, Rocks and Fossils.* 2nd ed. rev. and enl. New York: Cambridge University Press, 1999.
2. Bowes, Donald R., ed. *The Encyclopedia of Igneous and Metamorphic Petrology.* New York: Van Nostrand Reinhold, 1989.
2. Dasch, E. Julius, ed. *Macmillan Encyclopedia of Earth Sciences.* 2 vols. New York: Macmillan, 1996.
2. Good, Gregory A., ed. *Sciences of the Earth: An Encyclopedia of Events, People, and Phenomena.* 2 vols. New York: Garland, 1998.
2. Nierenberg, William A., ed. *Encyclopedia of Earth System Science.* San Diego: Academic Press, 1992.
2. Seyfert, Carl K., ed. *The Encyclopedia of Structural Geology and Plate Tectonics.* New York: Van Nostrand Reinhold, 1987.
2. Singer, Ronald, ed. *Encyclopedia of Paleontology.* 2 vols. Chicago: Fitzroy Dearborn, 1999.
4. *Bibliography and Index of Geography.* Alexandria, Va.: American Geological Institute. Also available online as *GeoRef* at http://georef.cos.com and as a CD-ROM.
4. *Geobase.* Norwich, Eng.: Elsevier. Available online from multiple sources and as a CD-ROM.
4. *ISI Web of Science: Science Citation Index.* Philadelphia: Institute for Scientific Information. Also available online at http://isio.isiknowledge.com.

4. Wood, David N., Joan E. Hardy, and Anthony P. Harvey. *Information Sources in the Earth Sciences.* 2nd ed. New York: Bowker-Saur, 1989.

5. Bates, Robert L., Marla D. Adkins-Heljeson, and Rex C. Buchanan, eds. *Geowriting: A Guide to Writing, Editing, and Printing in Earth Science.* 5th ed. Alexandria, Va.: American Geological Institute, 1995.

5. Dunn, J., et al. *Organization and Content of a Typical Geologic Report.* Rev. ed. Arvada, Colo.: American Institute of Professional Geologists, 1993.

Mathematics

1. Borowski, E. J., and J. M. Borwein et al., eds. *HarperCollins Dictionary of Mathematics.* New York: HarperCollins, 1991.

1. James, Robert Clarke. *Mathematics Dictionary.* 5th ed. New York: Van Nostrand Reinhold, 1992.

1. Schwartzman, Steven. *The Words of Mathematics: An Etymological Dictionary of Mathematical Terms Used in English.* Washington, D.C.: Mathematical Association of America, 1994.

2. Ito, Kiyosi, ed. *Encyclopedic Dictionary of Mathematics.* 2nd ed. 2 vols. Cambridge: MIT Press, 1993.

3. Pemberton, John E. *How to Find Out in Mathematics.* 2nd rev. ed. Oxford: Pergamon, 1969.

4. *East European Scientific Abstracts.* Arlington: JPRS.

4. *Mathematical Reviews: 50th Anniversary Celebration.* Providence: American Mathematical Society, 1990.

4. *Mathsci.* Providence: American Mathematical Society. Also available online at http://e-math.ams.org/mathscinet and as a CD-ROM.

5. *A Manual for Authors of Mathematical Papers.* Rev. ed. Providence: American Mathematical Society, 1990.

Physics

1. *A Dictionary of Physics.* Oxford: Oxford University Press, 2000.

1. *Dictionary of Pure and Applied Physics.* Boca Raton: CRC Press, 2001.

1. Sube, Ralf. *Dictionary, Physics Basic Terms: English-German.* Berlin: A. Hatier, 1994.

1. Thewlis, James. *Concise Dictionary of Physics and Related Subjects.* 2nd ed. Rev. and enl. New York: Pergamon, 1979.

2. *The Cambridge Handbook of Physics Formulas.* Cambridge: Cambridge University Press, 2000.

2. Lerner, Rita G., and George L. Trigg, eds. *The Encyclopedia of Physics.* 2nd ed. New York: UCH, 1991.

2. Meyers, Robert A., ed. *Encyclopedia of Modern Physics.* San Diego: Academic Press, 1990.

2. Parker, Sybil P., ed. *McGraw-Hill Encyclopedia of Physics.* 2nd ed. New York: McGraw-Hill, 1993.

2. Trigg, George L., et al., eds. *Encyclopedia of Applied Physics.* New York: VCH, 1996.

3. Shaw, Dennis F. *Information Sources in Physics.* 3rd ed. New Jersey: Bowker-Saur, 1994.

4. *American Institute of Physics Journals.* College Park, Penn.: AIP. Also available online at http://www.aip.org/ojs/service.html.

4. *Applied Science and Technology Index.* New York: H. W. Wilson. Also available online at http://hwwilsonweb.com and as a CD-ROM.

4. Bohme, Siegfried. *Astronomy and Astrophysics Abstracts.* Berlin: Springer-Verlag.

4. *Current Physics Index.* Westbury, N.Y.: AIP. Also available online at http://ojps.aip.org/spinweb.

4. *IEEE Xplore.* New York: IEEE. Also available online at http://ieeexplore.ieee.org/Xplore/DynWel.jsp.

4. *Inspec.* Edison: Institution of Electrical Engineers. Also available online at http://www.iee.org/Publish/INSPEC and as a CD-ROM.

4. *Institute of Physics Electronic Journals.* London: IOP. Also available online at http://www.iop.org/EJ.

4. *ISI Web of Science: Science Citation Index.* Philadelphia: Institute for Scientific Information. Also available online at http://isio.isiknowledge.com.

4. *Physics Abstracts.* London: Institute of Electrical Engineers.

5. Katz, Michael J. *Elements of the Scientific Paper: A Step-by-Step Guide for Students and Professionals.* New Haven: Yale University Press, 1985.

6. American Institute of Physics. *AIP Style Manual.* 4th ed. Westbury, N.Y.: AIP, 1990.

A Note on Some of Our Sources

We intend this bibliographical survey for those who find the rhetoric of research itself interesting enough to explore as a research problem. We based our choices on the two points that we have emphasized throughout the book.

First, we selected sources that reflect our belief that writing is not just the last stage of the project, but is from the beginning a means of discovery and critical thinking. That is a view commonly held in writing studies today. But we concentrate on an aspect of writing that the common view has too often ignored, even rejected: rather than treat the standard forms of discourse and style as constraining and coercive, we believe that they are in fact creative; they can motivate thinking, even discovery.

Second, we have shifted attention from the lone writer as the prime creative force to focus on the interaction of writer and reader and on how that interaction can help you not just to draft your work, but develop and test your argument, even help you with your research. We believe that some of the most creative moments of research occur not when we decide what *we* want to put in our report, but when we think about what *our readers* must see there if they are to read it well and trust its conclusions. Because most standard works on doing and reporting research tend to ignore those two issues, we do not cite those works here.

Inevitably, some researchers will think that we have ignored

or overlooked crucial names or books. But the study of rhetoric—
what in this book we might have called the "craft of engagement
between writers and readers"—leads into every human science.
The names and titles we could have cited are thus endless. We
cite here only the sources that we have used directly or that have
particularly interested one or more of the three of us. To that
degree, this is an idiosyncratic list.

GENERAL BACKGROUND

Almost every contestable issue in rhetoric begins with Plato's
Phaedrus and *Gorgias* (*Gorgias/Plato*, trans. Robin Waterfield [Ox-
ford University Press, 1994]) and Aristotle's *Rhetoric* (*On Rhetoric: A Theory of Civic Discourse*, trans. George Kennedy [Oxford
University Press, 1991]). The best discussion of what rhetoric is
for is Eugene Garver's *Aristotle's Rhetoric: An Art of Character*
(University of Chicago Press, 1994). Following Aristotle is Cicero's *De Oratore*, trans. J. S. Watson (Southern Illinois Press,
1986); and *De Inventione*, trans. H. M. Hubbell (Harvard University Press, 1976); and Quintilian's *Institutiones oratoriae*, ed.
James J. Murphy (Southern Illinois University Press, 1987). A
study that traces the classical tradition into the modern world is
Thomas M. Conley's *Rhetoric in the European Tradition* (University of Chicago Press, 1994).

The modern tradition begins with eighteenth-century rhetori-
cians such as George Campbell, *The Philosophy of Rhetoric*, ed.
Lloyd F. Bitzer (Southern Illinois University Press, 1963, 1988).
In the twentieth century, classic works include I. A. Richards's
The Philosophy of Rhetoric (1935; reissued by Routledge Press,
2002); Chaim Perelman and Lucie Olbrechts-Tyteca's *The New
Rhetoric: A Treatise on Argumentation*, trans. John Wilkinson and
Purell Weaver (Notre Dame University Press, 1969; originally in
French, 1958); Kenneth Burke's *A Grammar of Motives* (1945) and
A Rhetoric of Motives (1950) (both reprinted by the University of
California Press, 1969); and Wayne Booth's *Modern Dogma and
the Rhetoric of Assent* (Notre Dame Press, 1974). Some would in-
clude in the contemporary tradition the work of Jacques Derrida,

as found in *Margins of Philosophy*, trans. Alan Bass (University of Chicago Press, 1982).

Excerpts from across the tradition are in Patricia Bizzell and Bruce Herzberg's anthology, *The Rhetorical Tradition: Readings from Classical Times to the Present* (Bedford Books, 1990). A useful anthology of articles is *Essays on Classical Rhetoric and Modern Discourse*, ed. Robert J. Connors, Lisa S. Ede, and Andrea A. Lunsford (Southern Illinois University Press, 1984). A widely used textbook that interprets the classical tradition for today's writing student is Edward P. J. Corbett's *Classical Rhetoric for the Modern Student*, 3rd ed. (Oxford University Press, 1990). A survey of modern rhetoricians with a good bibliography is Sonja K. Foss, Karen A. Foss, and Robert Trapp's *Contemporary Perspectives on Rhetoric* (Waveland Press, 1985). A first-class survey of the history of rhetoric and its manifold topics is the *Encyclopedia of Rhetoric*, ed. Thomas O. Sloane (Oxford, 2001).

RESEARCHERS AND READERS

Rhetorical studies have always necessarily considered audiences, but only in the last half-century have they focused on particular social or disciplinary or academic contexts, especially on how communities of researchers differ not only in their common knowledge and beliefs, but also in the way their research sites and practices influence their discourse. To appreciate the extent of that explosion of interest in diverse "rhetorics" of different fields, you might go to the Library of Congress online and call up "the rhetoric of . . . ," filling in your field of special interest. Since about 1950, more than six hundred titles have emerged relating rhetorical study to this or that academic discipline, including sophisticated studies such as Donald McCloskey's *The Rhetoric of Economics* (University of Wisconsin Press, 1985); Alan G. Gross's *The Rhetoric of Science* (Harvard University Press, 1990); and Austin Sarat and Thomas R. Kearns's *The Rhetoric of Law* (University of Michigan Press, 1994). A seminal inquiry into these matters is Bruno Latour's *Science in Action* (Harvard University Press, 1987). See also Greg Meyers, *Writing Biology* (University of Wisconsin Press, 1990); and Charles Bazerman, *Shaping Written Knowledge* (University of Wisconsin Press, 1988).

Two useful anthologies of modern studies on particular readers are *The Rhetorical Turn: Invention and Persuasion in the Conduct of Inquiry,* ed. Herbert W. Simons (University of Chicago Press, 1990); and *Textual Dynamics and the Professions,* ed. Charles Bazerman and James Paradis (University of Wisconsin Press, 1991). Some research on the role of social forces has focused on gender: see Evelyn Fox Keller, *Reflections on Gender and Science* (Yale University Press, 1985); and a collection, *Body Politics: Women and the Discourses of Science,* ed. Mary Jacobus, Evelyn Fox Keller, and Sally Shuttleworth (Routledge, 1990). A penetrating analysis of how recent rhetorical theories resemble and differ from classical theories is given by James Kasteley in *Rethinking the Rhetorical Tradition* (Yale University Press, 1997).

ASKING QUESTIONS, FINDING ANSWERS

Almost all of the works we've mentioned so far are pertinent to the topic of "invention"—the pursuit of genuine new questions and finding new answers. These arts of inquiry begin with Aristotle's topics and Cicero's *De Inventione.* Among the most influential of the modern approaches is Richard Young, A. L. Becker, and Kenneth Pike's *Rhetoric: Discovery and Change* (Harcourt Brace Jovanovich, 1970). (The scheme of questions outlined in chapter 3 is based on Kenneth Pike's original work in tagmemics in the 1960s.) On the idea of "problem," see an old but still seminal book, John Dewey's *How We Think* (Heath, 1910). For a psychologist's point of view, see *The Nature of Creativity,* ed. R. J. Sternberg (Cambridge University Press, 1988). And on we could go, through every major field.

In the last few decades, there has been increasing interest in the psychological basis of reasoning, in particular in an area called "cognitive bias." A popular guide to cognitive bias is Stuart Sutherland's *Irrationality: Why We Don't Think Straight* (Rutgers University Press, 1992). Two more scholarly treatments of cognitive bias are Richard Nisbett and Lee Ross, *Human Inference:*

Strategies and Shortcomings of Social Judgment (Prentice Hall, 1980); and Jonathan Baron's *Thinking and Deciding,* 2nd ed. (Cambridge University Press, 1994).

PROBLEM POSING/SOLVING

The literature on problem solving is immense; the literature on problem finding is comparatively small. For general accounts of problem solving and finding, the classic source is John Dewey's *How We Think* (Heath, 1910). The most accessible contemporary work is that of Getzels and Csikszentmihalyi: J. W. Getzels, "Problem-Finding and the Inventiveness of Solutions," *Journal of Creative Behavior* 9 (1975); J. W. Getzels and Mihaly Csikszentmihalyi, *The Creative Vision: A Longitudinal Study of Problem Finding in Art* (Wiley, 1976); and J. W. Getzels, "The Problem of the Problem," *Question Framing and Response Consistency,* ed. Robin M. Hogarth (Jossey-Bass, 1982). Two studies of problems in particular fields are Murray S. Davis, "That's Interesting! Towards a Phenomenology of Sociology and a Sociology of Phenomenology," *Philosophy of the Social Sciences* 1 (1971) (which inspired our Quick Tip on contradictions); and Susan Peck MacDonald, "Problem Definition in Academic Writing," *College English* 49 (1987).

ARGUMENTS

Problems of argument, in one sense or another, have also permeated all of the books mentioned so far. Our own treatment of argument in this book began with Aristotle, but more particularly with Stephen Toulmin's *Uses of Argument* (Cambridge Press, 1958), a book that has changed the way many rhetoricians think about the formal structure of argument. His views were expanded in a textbook written with Richard Rieke and Allan Janik, *An Introduction to Reasoning,* 2nd ed. (Macmillan, 1984). A critique of Toulmin's approach with substantial bibliography is James B. Freeman's *Dialectics and the Macrostructure of Arguments* (Foris Publications, 1991). A good review of current work in argumentation in Canada and Europe can be found in Ralph H. Johnson, *Manifest Rationality: A Pragmatic Theory of Argument* (Erlbaum,

2000); and Frans H. van Eemeren, Rob Grootendorst, and Francisca Snoeck Hekemans et al., *Fundamentals of Argumentation Theory* (Erlbaum, 1996).

There is a long history of studying argument in more traditional ways. Extensive references are in Frans H. van Eemeren, Rob Grootendorst, and Tjark Kruiger's *Handbook of Argumentation Theory* (Foris, 1987). A useful application of conventional logic to argument is in David Kennedy's *The Art of Reasoning* (Norton, 1988). A textbook that addresses many aspects of written arguments is Williams and Colomb's *The Craft of Argument,* 2nd ed. (Addison-Wesley Longman, 2003).

There is a useful literature on argument as the basis for rationality. Among these works are Michael Billig, *Arguing and Thinking: A Rhetorical Approach to Social Psychology* (Cambridge University Press, 1987); George Meyerson, *Rhetoric, Reason and Society* (Sage Publications, 1994); James Crosswhite, *The Rhetoric of Reason: Writing and the Attractions of Argument* (University of Wisconsin Press, 1996); and most comprehensively, Richard Rorty, *Philosophy and the Mirror of Nature* (Princeton University Press, 1980). A journalistic view of today's obsession with destructive argument is Deborah Tannen's *The Argument Culture: Moving from Debate to Dialogue* (Random House, 1998). A provocative study of the role of emotion in rational thinking is Ronald de Sousa's *The Rationality of Emotion* (MIT Press, 1990).

DRAFTING AND REVISING

Detailed advice about organization and style is offered in *Style: Toward Clarity and Grace* (University of Chicago Press, 1990) by Williams, including two chapters coauthored with Colomb. A version limited to style but including exercises is Williams's *Style: Ten Lessons in Clarity and Grace,* 7th ed. (Addison-Wesley Longman, 2003). Two quite different ways to think about style are Richard Lanham's *Style: An Anti-Textbook* (Yale University Press, 1974); and Walker Gibson's *Tough, Sweet and Stuffy: An Essay on Modern American Prose Styles* (Indiana University Press, 1966). The classic works in the visual presentation of data are Edward

Tufte's *The Visual Display of Quantitative Information* (Graphics Press, 1983) and *Envisioning Information* (Graphics Press, 1990). Advanced students might look at William S. Cleveland's *Elements of Graphing Data* (Wadsworth Press, 1985), and his and Marilyn E. McGill's *Dynamic Graphics for Statistics* (Wadsworth, 1988). For the rhetoric of maps, see Mark Monmonier's *Mapping It Out: Expository Cartography for the Humanities and Social Sciences* (University of Chicago Press, 1993). An approach to introductions that also takes a structural view but offers a description usefully different from ours is in John Swales's *Genre Analysis: English in Academic and Research Settings* (Cambridge University Press, 1990).

ETHICS

The concern about the ethics of rhetorical transactions is as old as rhetoric itself. The two classical discussions are Plato's *Gorgias* and book XII of Quintilian's *Institutiones oratoriae*. The matter of rhetoric and ethics was revived in modern times by Burke's *A Grammar of Motives* and by Richard Weaver's *The Ethics of Rhetoric* (Henry Regnery, 1953), a book that still provokes controversy. A contemporary discussion of the more general notion of ethics in communication is Richard Johannesen's *Ethics in Human Communication*, 3rd ed. (Waveland, 1990). Recently, feminist scholars have critiqued the traditional view of argument as conflict in ways similar to ours, but based on the question of whether the standard forms of argument can ever be ethical, because they are, they claim, characteristically coercive and patriarchal. An influential, early version of this argument is Sally Miller Gearhart's "The Womanization of Rhetoric," *Women's Studies International Quarterly* 2 (1979): 195–201. For a discussion of why our culture predisposes us to think about argument as conflict, see George Lakoff and Mark Johnson, *Metaphors We Live By* (University of Chicago Press, 1980).

SOURCES OF FURTHER BIBLIOGRAPHY

An annual bibliography for research in teaching writing appears in the journal *Research in the Teaching of English*. An annual bibliography of rhetoric and composition was the *Longman Bibliography of Composition and Rhetoric*, ed. Erika Lindemann (Longman, 1987), now continued by *CCCC Bibliography of Composition and Rhetoric* (Southern Illinois University Press, 1990–). Journals that publish nontechnical articles on these topics include *College Composition and Communication*, *College English*, *Journal of Advanced Composition*, *Philosophy and Rhetoric*, *Pre/Text*, *Quarterly Journal of Speech*, *Rhetorica*, *Rhetoric and Public Affairs*, *Rhetoric Review*, and *Rhetoric Society Quarterly*. More technical work appears in *Applied Linguistics*, *Discourse Processes*, *Text*, and *Written Communication*. Because rhetoric is now conceived so broadly, look at citations in bibliographies of current articles for other journals to monitor.

Index

abstract, 194, 219–21; contexts of, 220; launching point in, 220; main point in, 219; problem in, 219; theme in, 219

abstraction, as character, 270–73

acknowledgment and response, 114, 118–19, 131, 151–64; placing, 199; selecting, 157–59; vocabulary of, 161–64

action, as noun, 267–69; as verb, 267

active voice, vs. passive, 275–77

alternatives, acknowledging, 157–58; counterexample, 156; definition, 156–57; predictable, 155–57

analysis, forms of, 91

anecdote, opening with, 239

answers, as claim, point, solution, 194

anxiety, 3, 6, 30, 80, 104, 126

argument, 114–181; complex, 121–22; as conversation, 11, 14, 17–18, 25, 40, 52, 56, 69, 112–15, 159, 208, 218, 227, 237, 287; evaluating, 217–18; finding alternative, 154–57; identifying, 216; model of, 91; organizing, 191–201; questioning, 152–54; revising, 216–18; thickening, 122; two reasons for failure, 124–25

Aristotle, 208

assignment, basing draft on, 191; characteristics of, 291–93; scenario, 292; scheduling, 293–94

audience. *See* reader

authority, 95–96, 146. *See also* expertise

background, 197. *See also* common ground

bar chart, 242–43; centrally divided, 257; constructing, 253–54; horizontal, 256–57; stacked, 255–56

benefit of solution, specifying, 230–31

bias, cognitive, 90–91

bibliographical guides, 41, 78, 80–81; experts as source, 86; footnotes as, 78; trails, 88

body of report, organizing, 196–200

catalog, library, 81–82

causes, alternative, 155–56

character, as abstraction, 270–73; defining, 270–72; as subject, 266–67

chart vs. graph, 250–51; introducing, 245–46; vs. table, 244; title of, 245–46

claim, 94, 114, 116–17, 127–37; as answer, point, solution, 194; conceptual, 129; contestable, 133–34; evaluating, 129; kind of, 127–29; language of, 129–30; logic of, 130–31; pragmatic, 129; qualifying, 135–37; significance of, 132–34; specificity, 129–31

clarity, 263–82; revising for, 281

collaborative writing, 26–31, 188

common ground, in introduction, 225–28; kinds, 227–28

commonplace, 165